OCEANOGRAPHIC ENVIRONMENT

Encyclopaedia of Environmental Pollution and Awareness in 21st Century Series

OCEANOGRAPHIC ENVIRONMENT

Edited by

V.K. Prabhakar

ANMOL PUBLICATIONS PVT. LTD.

NEW DELHI - 110 002 (INDIA)

ANMOL PUBLICATIONS PVT. LTD.
4374/4B, Ansari Road, Daryaganj
New Delhi - 110 002

Oceanographic Environment

First Edition, 2001

ISBN 81-261-0920-3

PRINTED IN INDIA

Published by J.L. Kumar for Anmol Publications Pvt. Ltd., New Delhi - 110 002 and Printed at Mehra Offset Press, Delhi.

Contents

PREFACE

Today, the cry of 'pollution' is heard from all the nooks and corners of the world, and pollution has become a major threat to the very existence of mankind on this earth. It is the major challenge of our times.

The pollution of various resources has gone to such an extent that we are unable to breathe fresh air and drink fresh water. On one hand the advancements of science and technology have added to the human comforts by giving us automobiles, electrical appliances, supersonic jets, space crafts, better medicines, better chemicals to control harmful insects and other pests, etc., but, on the other hand, they have given us a very serious problem to face that is pollution. Pollution can be defined as an undesirable change in the physical, chemical, or biological characteristics of the air, water, or land that can harmfully affect health, survival, or activities of humans or other living organisms.

Pollution does not have to cause physical harm, but it may merely interfere with human activities. Chemicals spewed into the air or water from an industrial plant may be harmful to humans and other organisms living near the plant. If expensive pollution controls are required, the plant may be forced to shut down.

India faces serious problems of environmental pollution. Water pollution from untreated community and industrial wastes flowing into our rivers and streams, increasing use of fertilizers and pesticides, dumping of organic wastes etc., are issues that need to be tackled on the priority. Air pollution, long regarded as the bane of industrialised nations looms large over our country.

Here, we are presenting this *Encyclopaedia of Environmental Pollution and Awareness in Twenty-first Century* in fifty volumes with a view to make people conscious about the seriousness of pollution in the present time and its remedies.

Editor

1 Introduction

The cycle of life in the sea, like that on land, is fueled by the sun's visible light acting on green plants. Of every million photons of sunlight reaching the earth's surface, some 90 enter into the net production of basic food. Perhaps so of the 90 contribute to the growth of the land plants and about 40 to the growth of the single celled green plants of the sea, the photoplankton. It is this minute fraction of the suns radiant energy that supplies the living organisms of this planet not energy that supplies the living organisms of this planet not only with their food but also with a breathable atmosphere.

The terrestrial and marine plants and animals arose from the same sources, through similar evolutionary sequence and by the action of the same natural laws. In some important respects this imaginary conditions is not unlike that of the dominant, food web of the sea, where almost all marine like is sustained by microscopic plants and near microscopic herbivore 45 and carnivorous, which pass on only a greatly diminished supply of food to sustain the larger, more active and more complex creatures.

In other respects the analogy is substantially inaccurate, because the primary marine food production is carried out by cells dispersed widely in a dense fluid medium this fact of an initial dispersal imposes a set of profound general conditions on all forms of life in the sea. In moderately rich areas of the sea, food is 100th times more dilute ii: volume and hundreds of thousands of time more dilute in relative mass. To crop this meager both a blind herbivòre or a simple pore in a filtering structure would need to process a weight of water hundreds of thousands of times the weight of the cell if eventually captures. In even the densest concentrations the factor exceeds several thousands and with each further step in the food web dilution increases. Thus from the beginnings of the marine food web we see many adaptations accommodating to this dilution: eyes in microscopic herbivorous animals.

Filters of exquisite design, mechanisms and behaviours for discovering local concentrations, complex search gear and, on the bottom. Attachments to elicit the aid of moving water in carrying out the task of filtration. All these adaptations stem from the conditions that limit plant life in the open sea to microscopic dimensions. It is in the sunlight near surface of the open sea that the unique nature of the dominant system of marine life is irrevocably molded. The near surface, or mixed, layer of the sea varies in thickness from tens of feet to hundreds depending on the nature of the general circulation, mixing by winds and heating. Gear the basic food production of the sea is accomplished by single called plants.

The organisms of the sea are bore, live breathe feed, excrete, move, grow, mate, reproduce, and die within a single interconnected medium. This interactions among the marine organisms and interactions of the organisms with the chemical and physical processes of the sea range across the entire spectrum from simple, adamant constraints to complex effects of many subtle interactions. A general discussions of a living system should consider the ways in which plants elaborate basic organic material from inorganic substances and the successive and often highly intricate stems by which organisms then return this material to the inorganic reservoir. The discussion should also slow the forms of life by which such processes are conducted. Some organic material is carried to the sea by rivers and some is manufactured in shallow water by attached plants. More than 90% of the basic organic material that fuels and builds the life in the sea, however, is synthesized within the lighted surface layers of open water by the many varieties of phytoplankton. These sunny pastures of plant cells are grazed by the herbivorous zooplanketon and by some small fishes. These in turn are prey to various carnivorous creatures, large and small, who have their predators also.

The debris from the activities in the surface layers settles into the dimly lighted and unlighted midlayers of the sea, the twilight mesopelasic zone and the mid night bathypelagic zone, to serve as one source of food for their strange inhabitants. This process depletes the surface layers of some food and particularly of vital plant nutrients, or fertilizers, that become trapped below the surface layers, where they are unavailable to the plants.

Food and nutrients are also actively carried downward from the surface by vertically migrating animals. The depleted remnants of this constant "rain" of detritus continue to the sea floor and support those animals that live just above the bottom, on the bottom and burrowed into the bottom. There filter feeding and burrowing animals and bacteria rework the remaining refractory particles. The more active animals also find repast in mid water creatures and in the occasional falls on carcasses and other larger debris. Except in unusual small areas there is an abundance of oxygen in the deep water. And the solid bottom presents advantages that allow the support of a denser populations of larger creatures than can exist in deep mid water.

In shallower water such as banks, atolls, continental shelves and shallows seas conditions associated with a solid bottom and other regional modifications of general regime enable rich populations to develop. Such areas constitute about 7% of the total area of the ocean. In some of these regions added food results from the growth of larger fixed plants and from land drainage.

One common group of small phytoplankton are the coccolithophores, with calcareous plants. A swimming ability and often an oil droplet for food storage and bouyancy. The larger microscopic phytoplankton are composed of many species belonging to several groups: naked algal cells, diatoms with complex shells of silica and activity swimming and rotating flagellates.

Very small forms of many groups are also abundant and collectively are called nannoplankton. The species composition of the phytoplankton is everywhere complex and varies from place to place, season to season and year to year. The various regions of the ocean are typified, however, by dominant major groups and particular species, seasonal effects are often strong, with dense blooms of phytoplankton occurring when high levels of plant nutrients suddenly become usable or available, such as in high latitudes in spring or along coasts at the on set of upwelling. The concentrations of phytoplankton varies on all dimensional scales, even down to small patches.

In addition the large scale circulation of the ocean continuously sweeps the pelagic plants out of the region to which they are best adapted. It is essential that some individuals be returned to renew the populations. More mechanisms for this essential return exist for single celled plants than exist for large plants or even for any conventional spores, seeds or juveniles. Any of these can be carried by oveanec gyres or diffused by large scale motions of surface eddies and periodic counter flow, but single celled plants can also ride submerged counter currents while temporarily feeding on the food particles or perhaps dissolved organic material. Other mechanisms of distribution undoubtedly are also occasionally important. For example, living marine plant cells are carried by storm-borne spray. In bird feathers and by wellfed fish and birds in their undigested food.

No large plant has solved the many problems of development, dispersla, and reproduction. There are no pelagic trees, and these several factors in concert therefore restrict the open sea in a profound way. They confine it to an initial food web composed of microscopic forms, whereas larger plants live attached only to shallow bottoms.

Attached plants, unlike free-floating plants, are not subject to the afore mentioned limitations. For attached plants all degrees of water motion enhance the exchange of nutrients and wastes. Although species of phytoplankton will populate only regions with conditions to which they are adapted, factors other than temperature, nutrients and light levels undoubtedly are important in determining the species composition of phytoplankton populations.

Little is understood of the mechanisms that give rise to an abundance of particular species under certain conditions. Grazing herbivores may consume only a parts of the size range of cells. Allowing certain sizes and types to dominate temporarily. Little is understood of the mechanisms that give rise to an abundance of particular species under certain conditions. Chemical by products of certain species probably exclude certain other species. Often details of individual cell behavior are probably also important in the introduction and success of a species in a particular area. In some cases we can glimpse what the mechanisms are for example, both the larger diatoms and the larger flagellates can more at appreciable velocities through the water.

The diatoms commonly sink downward, whereas the flagellates actively swim upward toward light. These are probably patterns of behaviour primarily for increasing exchange. But the interaction of such undirectional motions with random turbulence or systemic convective motions is not simple, as it is with an in active particle. Rather, we would expect diatoms to be statistically abundant in upward moving water and to sink out of the near surface layers when turbulence or upward convection is low.

Flagellates should be statistically more abundant in down welling water and should concentrate near the surface in low turbulence and slow downward water motions. These effects seem to exist. Of some continental costs in summer flagellates may eventually collect in high concentrations. As they begin to shade one to another from the light.

Each individual struggles close to the lighted surface, producing such a high density that large areas of the water are turned red or brown by their pigments. The concentration of flagellates in these "Red Tides" sometimes becomes too great for their own survival. Several specia of flagellates also become highly foxic as they grow older. Thus they sometimes both produce and participate in a mass death of fish and invertebrates that has been on own to give rise to such a high yield of hydrogen sulfide as to blacken the white houses of coastal cities. Large diatom cells, on the other hand, spread a disproportionately greater time in upward moving regions of the water and an unlimited time in any region where the upward motion about equals their own downward motion.

Diatom cells are there statistically abundant in upwelling water, and the distribution of diatoms probably is often are flection of the turbulant convective region of the water. Sinking and the dependence of the larger diatoms on upward convection and turbulence for support aids item in reaching upwelling regions, where nutrients are high, it helps to explain their dominance in such regions and such other features of their distributions as their high proportion in rich ocean regions and their frequent inverse occurrence with flagellates. Differences in adaptations to the physical and chemical products, probably reinforce such relations.

In some areas, such as parts of the equatorial current system and shallow seas. Where lateral and vertical circulation is rapid, the species composition of phytoplankton is perhaps more simply a result of the inherent ability of the species to grow, survive and reproduce under the local conditions of temperature, light, nutrients, competitors and herbivores. Elsewhere second-order effects of the detailed cell behaviour often dominate. These details of behaviour that give rise to concentration on any dimensional scale are particularly important to all subsequent steps in the food chain.

The grazing of the phytoplankton it principally conducted by the herbivorous members of that zooplankton. A heterogeneous group of small animals that carry out several steps in the food web as herbivores, carnivores and detrital feeders. Among the important members of the zooplankton are the arthopods, animals and external skeletons that belong to the same broad groups as insects, crabs and shrimps.

The planktonic arthopods induce the abundant copepods, which are in a sense the maring equivalent of insects. Copepods are represented in the sea by some 10,000 or more species that act not only as herbivores.

Carnivores or detrital feeders but also as external or even internal parasites' two or three thousand of these species live in the open sea. Other important arthropods are the shrimp like euphausidis, the strongest vertical migrators of the zooplankton. They compose the vast shoals of krill that occur in high latitudes and that constitute one of the principal food of the baleen whales. The zooplankton also include the strange bristle jawed chateognaths or arrow worms, carnivores or mysterious origin and affinities know only in the marine environment.

Widely distributed and abundant. The chateognaths are represented by a surprisingly small number of species. Perhaps fewer than 50. Larvae of many types, worms, medusae, ctenophores, gastropods, pteropods and heteropods, salps, unpigmented flagellates and many others are also important components of this milieu. Each with its own remarkably complex and often bizarre life history, behaviour and form.

The larger zooplankton are mainly carnivores, and those of herbivores habit are restricted to feeding on the larger plant cells. Much of the blood supply, however, exists in the form of very small particles such as the nanoplankton, and these appear to be available almost solely to microscopic creatures. Eyesight has developed in many minute animals to make possible selective capture. A variety of vebs, bristles, rakes, combs, cilia and other structures are found, and they are often sticky. Stickiness allows the capture of food that is finer than the interspaces in the filtering structures, and it greatly reduces the expenditure of energy.

A few groups have developed extremely fine and apparently quite effective nets. One group that has accomplished this is the "Larvacea". A larvacian produces and inhabits a complex external"House," much larger than its owner. That contains a system of very finely constructed nets through which the creature maintains a gentle flow. The larvacea have apparently solved the problem of energy loss in filtering by having proportionately large nets; fine strong threads and a low rate of flow. The composition of zooplankton differs from place to place, day to night, season and year to year, yet most species are limited in distribution , and the members of planktonic communities commonly show a rather stable representation of the modes of life. The zooplankton are, of course, faces with the necessity of maintaining breeding assemblages and like the phytoplankton.

With the necessity of establishing are inoculation of parent waters. In addition, their behaviour must lead to a correspondence with their food and to the pattern of large scale and small scale spottiness already imposed on the marine relean by the phytoplankton. The swimming powers of the larger zooplankton are quite adequate for finding local small scale patcho of food. That this task is accomplished on a large scale is indirectly demonstrated by the observed correspondence between the quantities of zooplankton and the plant nutrients in the surface waters. There are many large and small puzzles in the distributions of zooplankton.

As an ex-dense concentrations of phytoplankton are often associated with low populations of zooplankton. These are probably rapidly growing blooms that zooplankton have not yet invaded and grazed on, but it is not completely clear that this is so. Chemical repulsion may be involved. The concentration of larger zooplankton and small fishes in the surface layers is much greater at night than during the day, because of a group of strong swimming members that share their time between the surface and the mesopelagic region. Many small zooplankton organisms also make a daily migration of some vertical extent.

In addition to its primary purpose daily vertical migration undoubtedly served the (migration) migrating organisms in a number of other ways. It enables the creatures to adjust their mean temperature. So that by spending the days in cooler water the amount of food used during rest is reduced. Perhaps such processes as the rate of egg development are also controlled by these tactics.

Convincing arguments have also been presented to show that vertical migration serves to maintain species so that they will be more successful under many more conditions than if they lived solely in the surface layers. This migration must also play an important part in the distribution of many species. Interaction of the daily migrants with the water motion produced by daily land sea breeze alternation can hold the migrants off store by a kind of "Rectification" of the oscillating water motion.

More generally descent into the (water) lower layers increases the influence of submerged counter currents. There by enhancing the opportunity to regions and hence to find high nutrient levels and associated high phytoplankton productivity. Even minor details of behaviour may strongly contribute to success. Migrants spend the day at a depth corresponding to relatively constant low light, where the moment of water commonly is different from that at the surface. Most of the members rise some what even on the passage of cloud shadow. The principal food supplies of the pelagic populations are passed on in incremental steps and rapidly depleted quantity to the larger carnivores, zooplankton, then to small fishes and squids, and ultimately to the wide range of larger carnivores of the pelagic realm.

The Pelagic region contains some of the largest and most superbly designed creatures ever to inhabit this earth: the exquisitely constructed pelasic tunas. The multicolored dolphin fishes, captures of flying fishes; the conversational purposes; the shallow and deep feeding swordfishes and toothed whales, and the greatest carnivores of all, the baleen whales and some plankton eating sharks. Whose prey are entire schools of krill or small fishes. Seals and sealions feed far into the pelagic realm. In concert with these great predators, large carnivorous sharks a wait injured prey.

Marine birds, some adapted to almost continuous pelagic life, consume surprising quantities of ocean food, diving, plunging, skimming and gulping in pursuit. These larger creatures of the sea commonly move in schools, shoals, and herds. In addition to meeting the needs of mating such grouping is advantageous in both defensive and predatory strategy.

Small fishes of several species commonly school together. Diverse predators also form loosely cooperative groups, and many species of marine birds depend almost wholly on prey driven to the surface by submerged predators. At night, schools of prey and predators are almost always spectacularly illuminated by bioluminescence produces by the microscopic and larger plankton.

The reason for the ubiquitous production of light by the micro organisms of the sea remains obscure, and suggested explanations are controversial. It has been suggested that light is a kind of inadvertent by product of life in transparent organisms. It has also been hypothesized that the emission of light on disturbance is advantageous to the plankton in making the predators of the plankton conspicuous to their predators. The fall out of organic material into the deep dimly lighted mid water supports a sparse population of fishes and invertebrates with in the mesopelasic and bathypelagic zones are found some of the most curious and bizarre creatures of this earth. These range from the highly developed and powerfully predators intruders, toothed whales and sword fishes, at the climax of the food chain to the remarkable squids, octopuses, euphausidis, lantern fishes, gulpers and angler fishes that in habit the bathypelagic region. In the mesopelagic region, where some sunlight penetrates, fishes are often countershaded, that is, they are darker above and lighter below. As are surface fishes. Many of the creatures of this dimly lighted region participate in the daily migration. Swimming to the upper layers at evening like bath emerging from their caves. There are some much larger stronger and more active fishes and squids in this region, although they are not taken in trawls or seen from submersibles.

Knowledge of their evidence comes mainly from specimens found in the stomach of sperm whales and sword fish. There is evidence that the sperm whales posses highly developed long range hunting sonar. They may locate their prey over relatively great distances. Perhaps miles, from just such an extremely sparse population of active bathypelagic animals.

Although many near surface organisms are luminescent. It is the bathypelagic region that bioluminescence has reached a surprising level of development. With at least two thirds of the species producing light. Clearly bioluminescence can be valuable to higher organisms, and the creatures of the bathypelagic realm have developed light producing organs and structures to a high degree. Strong fishes may confuse predators by "target alteration" effects or by producing residual images in the predators vision. Some squids and shrimps are more direct and discharge luminous organs are arranged on some fishes so that they can be used to countershade their siltouettes against faint light coming from the surface. Lights may also be used for locating a mate. A problem of this vast, sparsely populated domain that has been solved by some angler fishes by the development of tiny males that live parasitically attached to their relatively huge mates. Light level in the bathypelagic regain can be much lower. This is most probably the primary difference that accounts for the absence of bioluminescence in higher land animals and the richness of II's development in the ocean forms.

There are some 2000 species of the larger invertibrates known to inhabit the bathypelagic zone. But only 4 few of these species appear to be widespread. The barriers to distribution in this widely interconnected mid water region are not obvious. The floor of the deep sea constitutes an environment quite unlike the midwater and surface environments. There are sites for the attachment of the larger invertebrates that filter detritus from the water. Among these animals are representative of some of the earliest multicelled creature to exist on the earth glass sponges, sea littles—once thought to have been long extinct—and lamp shells.

Larger numbers of active fishes and other creatures are attracted to the bait almost immediately. It is probably I feel that several rather independent branches of the food web coexist in support of the deep bottom creatures: one the familiar rain of fine detitus, and the other the rare, widely separate falls of large food particles that are in excess of the local feeding capacity of broadly diffuse bathypelagic population.

Such falls would include dead whales, large sharks or other large fishes and fragments of these, the multitude of remnants that are left when predators attack a school of surface fish and undoubtedly, garbage from ships and kills from underwater explosions. These sources result in an influx of high grade food to the sea floor, and we would expect to find a population of active creature adapted to its prompt discovery and utilization. Other sources of food materials are braided into these two extremes of the abyssal food web.

There is the rather subtle downward diffusion of living and dead food that results initially from the daily vertical migration of small fishes and zooplankton near the surface. This migration appears to impress a sympathetic daily migration on the mid water population down to great depths, far below the levels that light penetrates. Not only may such vertical migration bring feeble bathypelagic creatures near the bottom but also it accelerates in it self the flux of dead food material to the bottom of the deep sea.

Benthic animals are much more abundant in the shallower mater off continents, particularly offshore from large rivers. Here there is often not only a richer near surface production and less hazardous journey of food to the sea floor but also a considerable input of food conveyed by rivers to the bottom. The deep slopes of river sediments wedges are typified by a comparatively rich population of burrowing and filtering animal that utilize this fine organic material.

The shallow regions of such wedges are highly productive of active and often valuable marine organism. The bottom is much more variable than the mid water zone is. There are as a result more environmental riches for an organism to occupy, and hence we see organisms that are of a wider range of from and habit. Aside from the wide range of form and fraction the benthic environment elicits from its inhabitants, there are more fundamental conditions that influence the nature and form of life there.

2 OCEANOGRAPHIC ENVIRONMENT

Marine ecology is concerned with environmental factors and problems of organic adjustments quite different from those in fresh water. Animals are relatively more conspicuous than plants. Succession is less evident, but such ecological processes represented by chemical cycles, co-operation, and disoperation, food chains, productivity, population, dynamics, niche segregation, speciation, and dispersal are fully as important as on land.

Distinct self-contained community units are more difficult to recognize in the sea than on land because of the apparently greater interrelation of benthic species and the free movement with circulating currents of plankton and nekton. Plankton is everywhere a basic link in food chains, but the general distribution and importance of plankton species in the sea is no more remarkable than that of soil organisms in terrestrial biomes.

To consider the entire ocean community as a single biome, as has been suggested by some investigators, is stretching the concept beyond its usefulness. Since we identify biomes by differences in the life-forms, and functional adjustments of the conspicuous dominant or predominant organisms, we may properly recognize biomes that occur in the open ocean, on eroding rocky shores, on muddy and sandy beaches, and composing the coral reefs and atolls.

Habitat

Marine biocycle is considered to have benthic (bottom) and pelagic (open water) divisions, the littoral zone of the ocean shore exceeds between the limits of high and low tides, the sublitoral zone covers the, continental shelf to a depth of about 200 m. The approximate depth at which maximum wave action produces any effect. The average depth of the ocean is about 3800 m, but oceanic trenches (hadal zone) extend much deeper, the Marianas Trench in the Pacific Ocean to approximately 11,6000 m. The neritic biochare is above the continental shelf and is commonly 16-240 km. (10-150 mi) wide.

Tides

Tides are caused by the attraction of the moon and to a lesser, extent, the sun. When the sun's attraction is added to that of the moon, as occurs twice each month at the time of full moon and new moon, the fluctuations of the tides are unusually high and unusually low, these are called spring tides. When the tidal influences of sun and moon are opposed, as happens twice each month, the tides have the least amount of flow and ebb and are called neap tides. The level of water in the ocean rises and falls usually twice each day or at an interval of 12 hours and 26 minutes. In some parts of the world the tides are less regular or there may be but one daily. Flood-tide is the period in which the level is rising and covering more and more of the shoreline; ebb-tide is the period in which the water level is less than a meter, but the change may be much more than this on the shore, depending on its configuration, the Bay of Fundy opens broadly to the sea and tapers to a narrow head landward, and tides may be 6 to 10 or even 15 m.

On the other hand, when bodies of water have only a relatively narrow connection with the sea, as does the Gulf of Mexico with the Atlantic, the range in water level is less than 30 cm. Even lakes have seiches, but they are hardly perceptible except in the larger lakes, where it may amount to a few centimeters. Tides have their greatest effect on animals on the seashore, because of the associated pounding of waves and the alternative submergence in water and exposure to the air. However, the organisms appear well adjusted to this rhythmic submergence and exposure.

For instance, as the stones on which the chiton occurs become exposed, the animals react positively to gravity and negatively to strong light, and move downward. They travel at maximum speed while the stone is still moist and become aggregated on the damp lower sides of the stones. When the stone again becomes immersed by the returning tide, the animals lose their geotactic orientation, and since illumination becomes more or less equal on all sides of the stones, they move about at random until they reach the upper surfaces again.

Substratum

The pounding action of waves or rocky shore may have tremendous force, estimated in one instance at 15,000 kg/ cm^2. animals occupying exposed rocky shores in the surf belt must be strongly protected and firmly attached. The conical shaped limpets present a minimum of surface to the waves. Large depressions in the rock retain water at ebb tide to form tidal pools and thus may contain the more delicate species because of the protection they afford.

The various sea weeds absorb some of the wave shock for the animals living with them. The shape, form and size of corals, sponges, and other colonial type are affected by the amount of wave action to which the animals are exposed, a shell of the Mytilus mussel may weigh 58 g where the animal is exposed to a heavy surf but only 26.5 g in more quiet waters. Sandy beaches occur only where the force of waves is reduced by being spread over a more gentle slope. Even here, especially during storms the sand makes a very unstable substratum and not many animals except mollusks and some of the echinoderms can keep from being smothered or buried. Mud bottoms occur only in relatively quiet waters. Burrows made in mud hold their form better than in sand, so large population of animals can occur in mud. Organic deposits are either calcareous or siliceous, the former being derived from the shells of foraminiferans, small pelagic mollusks, or flogellate coccolithophorids and the latter from skeletal material of diatoms and radiolarian protozoans.

Pressure

There is an enormous increase in the pressure of water upon the bodies of animals at great depths, and this may effect the vertical distribution of such species as crabs and mussels. This is not, however, an important limiting factor in the vertical distribution of animals in general, as internal pressures closely counter- balance external pressures and life is known to exist at the greatest depths. Adjustments of internal pressures are not so rapid, however, to prevent injury in many species that are dredged at great depths and quickly hauled to the surface.

Temperature

The temperatures of surface waters vary between the freezing point (–1.9°C) in the tropics. Seasonal variations are small in polar and tropical waters but somewhat greater in the temperate zones. Temperature varies with depth, more so in the tropics than elsewhere. At 60°N latitude in the Atlantic Ocean, the mean temperature of the warmest and coldest months beneath the surface is about 10°C. On the equator the temperature beneath the surface is approximately 26°C, at 200 m, 13°C, at 1000 m 4.51°C, and at 2000 m 3.3°C (Ekman 1943). In temperate regions, seasonal thermoclinc develops near the surface in summer and is destroyed in autumn and winter when vertical mixing creates a layer of relatively uniform temperature in the upper 20 to 300 m.

Currents

Currents moving toward the poles from the equator consist of warm water, and currents moving in the opposite direction of cold water. Surface currents make wide circular movements in opposite directions in the northern and southern hemisphere.

Light

Much of the solar radiation is reflected from the surface of absorbed in the upper layers. Even in the clearest waters and at maximum radiation; the red, orange and ultraviolet are absorbed in the first 20 m. Green, yellow and blue wavelengths penetrate further, depending on the water colour. When the sun is not at the zenith, light penetration is reduced, and the maximum penetration in the winter at higher latitudes is much less than during the summer.

Salinity

The salinity of sea water varies from place to place depending largely on the amount that it is diluted by the yellow of fresh water from rivers or melting glaciers or the amount that it is concentrated by evaporation. The Red Sea, for instance has a salinity of 40% (40 g of dry salts in 1000 g of seawater) while in some polar seas the salinity is less then 30%. The average salinity of the oceans as a whole is commonly given as 35% of which the chlorideion constitutes about 19% and the sodium ion little over 10%.

The various major salts occur nearly everywhere in definite and constant proportions. As one would expect, the pH of seawater is high, averaging about 8, there is some similarity in relative proportions and concentrations of the variations in seawater and in the blood of body fluids of many invertebrate organisms. This may indicate that the sea is the habitat in which living forms first evolved. The contrast in salinity between seawater (35,000 ppm) and fresh water (15,660 ppm) requires important difference in physiological adjustment of organisms to occupy these two habitats. The problem is one of osmotic regulation.

Most marine invertebrates are poikilosmotic in that they are nearly isotonic with sea water, they are highly permeable to water and gain or lose water according to concentration of the medium. A few marine segmented worms, flatworms, and crabs and all marine fish and mammals have at least some internal osmotic regulation and tend to be homoisomotic.

All except the elasmobranch fishes maintain body fluids hypotonic to seawater in various ways. The skin has decreased its permeability to the free movement of water back and forth, the necessary water is obtained by swallowing surplus salts are secreted outside the body, especially through the gills, and there is a general decreased function or atrophy of water secreting organs such as the kidneys. The practical absence of insects and amphibians from the sea is largely due to their inability to secrete salts outwardly. The high osmotic concentration found in elasmobranches is the result of huge quantities of urea retained in the body tissues and fluids. These salts keep cycling through the ecosystem, but additions to the supply some come continually from the land, being washed into the oceans by the rivers.

Neritic waters are especially fertile and support a great mass and variety of animal life because of this land drainage and the pattern of water circulation on the continental shelf. Biological productivity decreases progressively from shallow waters over the continental shelf, to deeper waters, to the open ocean, but it also has high over offshore banks and in areas of upwelling substantial amounts of nitrogen salts are also swept out of the air by precipitation and there is nitrogen fixation by bacteria.

Oxygen

The oxygen supply of seawater comes by diffusion from the air at the surface and from photosynthesis of green plants down to the compensation point. It is continuously used at all depths in respiration of animals and plants and in the decomposition of organic matter. The oxygen content of seawater is seldom limiting for the occurrence of animals, except in the deeper waters of the brackish Black and Caspian Seas, where it is practically absent. Oxygen concentration is especially high on shores where there is splashing of waves.

Surface waters of the Atlantic Ocean commonly have 4.5 to 7.5 cc/liter and abyssal regions may run 5 cc/liter. Oxygen is somewhat less abundant in the Pacific and Indian Oceans. Oxygen may be reduced to lower concentrations between 100 to 1500 m, because of its use in animal respiration and in decomposition, that at lesser depths, where there is photosynthesis, or at greater depths, where the abundance of animals is greatly diminished.

Marine animals have a variety of mechanism and adaptations for respiration. Greatest difficulties occur in shore animals at low tide when they are exposed to the air, but the needs for oxygen at this time is decreased in many forms by curtailment of activity. Some crabs, barnacle, snails, and fish have become almost amphibious in being capable of respiring in air, although at reduced rates as well as in water. Pune mud bottoms may present anaerobic conditions a short distance below the surface, but mud bottoms mixed with sand contain an abundant and diversified form.

Plankton

The plankton of the sea includes a great variety of forms, even more than in fresh water. Rotifers, however, are uncommon in marine plankton and cladocerans are much less important. The nannoplankton consists mostly of protozoans, algae, bacteria, fungi, and viruses. The bacteria are largely periphytic, in that they are attached to the surfaces of floating plants, animals and to particles of detritus. Very few occur freely suspended in the water. Bacteria, including nitrifying forms, occur at all depths but are especially abundant in or close to the bottom. They are generally more numerous in the winter than in the summer. Nitrogen-fixing organisms are chiefly blue-green algae and possibly yeasts.

The green phytoplankton is composed primarily of diatoms, dinoflagellates, and small unarmored flagellates, but several other kinds of algae are present and occasionally important. The dianoflagellates Noctiluca and Ceratium are luminescent and in some regions may give a glow at night to the entire sea. Bioluminescene is not limited to these organisms, however, but occurs also in various forms of bacteria radiolarians, sponges, coelenterates, ctenophores, nemertineans, worms crustaceans, brittlestars, mollusks, balanoglossids, tunicates, and fish. The most important groups of protozoan zoo-plankton, other than the green flagellates which are usually considered with the phytoplankton, are the rhizopond foraminifera, the actinopod radiolaria, and the ciliate tintinnids. They may be enormously abundant at times.

The actual abundance of plankton varies greatly from place to place and from one season to the next. Smaller species tend to be more numerous than larger ones. The mean annual abundance of diatoms is commonly in the tens of thousands per litre and for shorter periods the year algal blooms may increase the population to hundreds of thousands of cell per liter, zooplankton is, however, much less abundant. Large, populations of zooplankton generally follow large populations of phytoplankton and by their grazing maintain the standing crop of phytoplankton at the size suited to prevailing conditions.

The total net zooplankton (cc per unit volume of water) is some 50 times more numerous in the neritic coastal waters of the Atlantic coast of North America than in the sargasso Sea. The abundance of plankton is generally higher in cold than in warm ocean waters correlated with the greater amount of phosphate present in colder water. Cold-water plankton tends to of be larger individuals size. There is generally in most taxonomic groups, however, a lower variety of species in cold than in warm waters. Annual productivity is also less in cold waters correlated with fewer generations per yaer. Productivity of zooplankton varies between 14 and 58 per cent of the net primary productivity.

Nekton

Mollusks, fishes, birds, and mammals make up the nekton of the sea. Mollusks are represented by the squids; fish by the sharks, flying fish herrings, meackerels, as many others, including numerous varieties of small species; and mammals, by the seals, tortoises, and whales The distribution of fish is irregular, but in general they occur more abundantly in neritic waters than in the open ocean. Likewise, they are much more numerous in the epipelagic than in lower strata. Most pelagic fish, except sharks, possess a swimbladder useful for maintaining hydrostatic equilibrium at the depth where they ocur; those fish that lack one are commonly bottom forms. in arctic waters fish are less abundant, and mammals relatively more important, than is the cause farther south. Birds, like many other marine animals, are more numerous in the neritic blochore than in the open ocean. In the ocean far from land occur only penguins, albatrosses, hearwaters, and petrels, and even these species become more common shoreward. Ohter marine species in neritic water are tropicbirds pelicans, gannets, boobies, cormorants, frigate-birds ducks, gulls terns skimmers asks and murres. These marine birds may spend many days of weeks feeding and travelling over the water, but all must search out some shore cliff, or isolated islands on which to nest. Here they sometimes concentrate in enormous numbers during the nesting season because of the limiting number or suitable nesting locations available.

Benthos

Benthos is of much greater variety in marine than in fresh water habitats. These animals are very abundant in the littoral zone and decrease in numbers with depth unit only scattered individuals are found in the deep ocean frenches. Benthos consists of sessile forms, the sponges barnacles, mussels, oysters, crinolds, corals, hydroids, bryozoans and some worms; creeping forms, such as carbs, lobsters, certain copepods, amphipods, other crustacoans, many protozoans snails, echinoderms, some bivalves, and some fishes; and burrowing forms, including must clams, worms, and some crustaceans. Sessile and creeping forms are often grouped as epifauna are burrowing forms as infauna.

Epifauna in the littoral zone decreases in variety towards the poles since it is subjected composition of infauna remains about the same. In food habits, different species may be grouped as seston eaters, mostly sessile or semi-sessile forms that capture suspended food particles; sluggish motile forms collecting detritus or food particles from the bottom surface; sluggish forms that extract food particles from bottom material that they swalloos; carnivores, and scavengers.

Plankton and Nekton Biome

This biome is characterized by the predominance of organisms possessing life-forms adopted to keep them afloat. Plankton and nekton predominate, although the deep-sea benthos may be considered as belonging to this biome. Seasonal aspection may bring drastic changes in species composition, especially in plankton. Dominance, in the sense used for terrestrial communities, probably does not exist, except possibly in the Sargasso Sea, where the floating vegetation establishes the habitat. The ecosystem is self-contained, however, since energy is derived from the sun and nutrient material continues to recirculate with little or no dependence on terrestrial resources. The sargassum community of the Atlantic Ocean is of special interest, on the floating sargassum alga accumulates and is held within a limited area by circular ocean currents. This plant belongs to the intertidal zone of the Caribean islands but in turn lose in large amounts along with attached animals during the hurricane season. It continues to grow thereafter, but does not reproduce. The fauna that it contains to a truly littoral one rather than pelagic, but because the algae to cold and ice erosion, but the species accumulates in fresh amounts as fast as old plants die, the animals reproduce and maintain a continued existence far from any shore.

Food Chains

As in fresh-water and terrestrial communities, bacteria in the sea are largely responsible for the final decomposition of excreta and dead bodies to make their essential nutrients available for reabsorption by the green phytoplankton. Nitrogen and phosphorus are least concentrated near the surface of the ocean, since this is the stratum in which they are most rapidly absorbed by the phytoplankton. Excreta and dead organisms sink during the process of decomposition, so nitrogen regeneration is most evident at depths of 500 to 1500 m. The total non-living organic matter, either dissolved or in the form of particulate matter, is generally much larger than the biomass of living organisms at any one time.

Organic matter makes up less than one-third of the total particulate matter, and all of it sinks slowly through the water, requiring months or even years to reach the bottom. Much of the particulate organic matter is still undissolved and accumulates on the ocean bottom. Numerous species of invertebrates depend on it for food, especially on the bacteria that it contains. The deep-sea fishes feed on these invertebrates or are carnivorous on other fish. Many of them have very wide mouths, distensible stomachs, and formidable teeth.

In addition to these food corrections it is also likely that many deep-sea fish and larger invertebrates undergo vertical migrations, so that they obtain food by preying on living organisms at more moderate depths. Much that is known about the life histories of these deep benthic species has been summarized by Marshall. Especially fertile regions of the open ocean occur when there is deep mixing of waters by turbulence and upwelling.

Vertical water currents bring nutrients up to the surface from intermediate depths where they had accumulated. Prominent regions of more or less permanent upwelling occur around the Antarctic continent, off the coasts of California, peru, and Somali, and off the west coasts of both north and south Africa. In many areas, very small phytoplankton and bacteria form the first link in the food chain, which goes next to the microzooplankton before reaching the carnivorous net zooplankton of copepods, euphasids, and other crustacean forms.

Most animals depending on small organisms and organic detritus for food have various filter-feeding mechanisms for straining the material out of the water. They do not actively search and catch individual items through directed actions. Invertebrate animals may also be able to absorb some essential salts and dissolved organic compounds to build their skeletal structures and for general metabolism, but there is considerable controversy on this point. The balcen or whalebone whales (Mysticeti) are toothless but possess large plates in their mouths that strain out the plankton (especially copepods, euphausiaceans, mysidaceans) that they use as food. Only occasionally are small fish as food. Only occasionally are small fish or other invertebrates ingested. Some whales reach tremendous proportions, and the differential in size between these animals and their food is one of the most remarkable in the animal kingdom.

Much more common is the feeding on plankton by squids, the young stages of most fishes, and such adult fishes as sardine, anchovy, menhaden, herring, and mackerel. The Menhaden is unique in having such fine-mesh gill-rakers that it can feed extensively on diatoms, which because of their smaller size cannot be readily secured by other large mariue animals.

Balanoid - Gastropod - Thallophyte Biome

This community extends form high-to below low tide levels on rocky shores. Benthic animals and attached algal plants are conspicuous and important, the benthos is mostly epifauna, as too hard to permit development of extensive infauna. When the tide is out the organisms are subjected to drying, the occasional inflow of fresh water, higher temperatures, and greater light intensities. Organisms to avoid desiccation when the tide is out by variously crawling under stones or thick algal growths, closing thick shells or operculae, retreating into crevices, or secreting a mucous seal.

Most organisms are also faced with the pounding action of waves, various holdfast or anchoring devices have developed, and many species protect their more delicate structures with a hard shell. The adaptations for life on the seashore are many and varied. The plankton and nekton associated with the benthon include many species not common to the oceanic biome.

Zonation

Vertical zonation of species on rocky shores is usually conspicuous, although individual species may extend widely into adjacent areas. Beginning on the landward side there is a supralittoral zone mostly above the action of tides and inhabited as much by land as by marine animals. This is followed seaward by a supralittoral or Littorin firnge which is wetted by the highest tides and by the splashing of waves. Because of the presence of either Myxophyceae or lichens, this zone is often discovered; commonly, black. The fringe is especially characterized by large numbers of small snails and sometimes isopods. Next below this fringe is the midlittoral or balanoid zone.

It is strictly intertidal being covered and uncovered every day, and is occupied characterically by acorn barnncles. This zone is often divided into subzones with the brancles predominate in the upper protion, while, polychaets, colonial hydroids, or other forms are relatively more important in the lower part. The subzonation of algae as often also well marked. The lowest zone over exposed, and then only at extreme low tides, is called the infralittoral fringe. It is a transition area. The entire area between extreme high and low tides, including the mid-littoral fringes, when considered as a unit may be referred to as littoral, eulittoral, or tidal zone to distinguish it from the infralittoral or sublittoral zone that extends from the lowest of low tides to the edge of the continental shelf.

Littoral Zone

Brown algae from thick masses and give protection to those animals that find shelter in or under them. A fauna of copepods, ostracods, water mites, and young littorinids inhabit these seaweeds. In England, the numbers of individuals per 100 g of seaweed vary from about 44 on brown algae to over 13,000 on lichens. The animal life on rocky shores is varied and luxuriant. Several species of acorn barnacles, snails, marine limpets, marine mussles, goose barnacles, sea anemones, chitons, sponges, hydroids, bryozoans, flatworms, annelids, isopods, crabs, sea urchins, starfishes, tunicates, and insects are present. Total abundance of animals may run into tens of thousands of individuals per square meter.

Sublittoral Zone

This community is not subjected to exposure by tides or to the pounding of surf but is affected considerably by wave action and the complete circulation of water. Animals move around somewhat more freely and there is less need for strong holdfast structures. Most organisms lack physiological tolerance for long exposure to the air and hence differ fundamentally in structure and more from the community described above. Laminarias or kelps are the largest of the brown algae and occur commonly in this community. They have root-like holdfasts attached to the bottom and their stalks, which are often several meters long, bear leaf-like branches that float at the surface in the larger species. A long list of animals find shelter and food in the kelp beds and especially in the protection of the holdfasts, polychaet worms are particualrly abundant in these holdfasts. Filamentous red algae (Rhodophyceae) are also prominent. Abundant characteristic animals on the Pacific coast are sea urchins, sea cucumber, starfishes, snails, rock oyster, chitons, limpets, scallops, mussels, nudibranches, barhacles, crabs, hermit crabs, hydroids, tunicates, shrimps, and various fish. Distribution of fish species correlates strongly with the type of bottom or benthos that is present.

Tidal Pools

Seawater is often retained as depressions or pools in the littoral zone and hence organisms here are never completely exposed to the air. They are however, subject to high light intensity an increase in the temperature between tides. Tidal pools are usually rich in both plant and animal life, and some species are largely restricted to them. Red algae and kelps perfer the more shaded, cooler pools; the green algae and some of the smaller brown algae predominate in the well-insolated pools. Animals of both the sublittoral and littoral zones are found here.

Pelecypod Annelid Biome

This biome develops on depositing sand and mud bottoms in contrast to the biome just described that occurs on eroding-rocky shores. There is still a good deal of wave action over sandy botoms. Fine sand particles shift about almost continuously, and animals have difficulty in preventing their burrows from coallapsing. In general, the water over muddy shores is shallower, quieter, and warmer. The mud forms a soft, comapct bottom, but is also easily moved or shifted around by storms and wave action. Animal burrows in mud are more permanent. Species tend to segregate depending on the amount of organic matter present. Shores of high mud content may be low in oxygen because of decaying organic matter, so animal populatins tend to be largest and most varied in a mixture of mud and sand. Tidal currents are weaker, and change in the level of water is less pronounced on sand and muddy shores than on rocky ones.

Important plants in this biome are the marine eelgrass, which is a seed plant and green algae, particularly the sealettuce, which grows in sheets either attached to the substratum or lying fragmented over large areas, and Enteromorpha, which grows in tufts or tangles. Occuring on eelagrass and sea-lettuce may be several kinds of epiphytic algae. These plans form extensive stands and are important to animals for attachement, shelter, and food.

Eelgrass was almost eliminated from the Atlantic coast in 1931-32, possibly becuse of a protozoan disease. This disturbance had a profoundly deleterious effect on the abundance of many animals, including the brant; a bird that depended on it almost exclusively for food, and on scallop and other coastal fisheries. Twenty years later there was evidence that eelgrass was recovering much of its former abundance.

Predominant animals are pelecypods, polychaete worms, particularly Arenicola and Nerels; starfishes, brittle-stars, sea cucumbers, crabs, amphipods, and snails. Populations may run to several thousands of individuals per square meter. A variety of small fish occur here. Birds inculde sandpipers, plovers and herons. The biome is worldwide in distribution, but the characteristic life-forms are represented by different species locally thus a number of secondary communities (Biociations) may be recongized.

Many of the animal constituents in this biome are burrowing forms. The sub-stratum of mud and sand holds considerable water and when the tide is out on exposed flats, pelecypods, worms and other animal constituents retract their fleshy organs into their burrows shells and remain in a water saturated environment. They are thus not exposed to the atmosphere with changes in the tide. Furthermore, most forms are generally tolerant of low oxygen and high carbon dioxide concentrations. In order to maintain respiration when retracted in their underground burrows these animals have long siphons, sometimes longer than their bodies, or long tubes or canals that extend to the surface. Through these they maintain a circulation of water, often .by means of special pumping organs.

Coral Reef Biome

Coral reef are formed by the accumulation of the calcreous skeletons of myriads of organisms. They extend from the sea bottom at depths of 46 m or rarely 74m. to slightly above low tide level. The best formation of coral reefs is confined to warm waters above 10°C, although individual species may extend into colder regions.

Predominant organisms involved are commonly the anthozoan stony corals and organ corals and the hydrozoan milliporids. Some reefs however are formed principally by forminifera and still others by clacareous algae. All massive coral structures employ calcareous algae as cement, these algae not only thrive in the pounding surf on the windward side of the reef, but by their growth are able to repair damage to the reef caused by storms.

Most typical reef-building animals are colonial and of shapes varying from closely compact, globose, or encrusting, to loosely branched or dendritic, depending in part on their exposure to wave action. Each polyp in a colony secretes its own clacareous skeleton and when it dies the next generation builds on top of the old so that the accumulation of a lime structure is fairly rapid. The bright yellow or red colors or corals near enough to their water's surface for adequate light penentration are the result of algae, the zooxanthellae, which are either embedded in the body wall or free in the internal cavities. In addition, there are bands of green filamentous algae growing to a depth of 2 or 3 cm in the pores or the insert coral skeleton that may have a biomass 16 times that of the zooxanthelate. In their photosynthesis these algae may absorb carbon dioxide and nutrients derived from animals of the coral reef and liberate oxygen of value to the animals. Perhaps because of this symbiotic relationship which requires solar radiation, living corals are largely confined to the upper, shallower waters. Coral animals actively ingest zooplankton, but apparently not phytoplankton, form the surrounding water. The three great biocycles-ocean, freshwater, and land-come into contact with each other around the margins of the seas. The change in the physical nature of the habitat from salt water to fresh water is a drastic one, but not more drastic than the change from slta water to land: The transition of animal and plant life is abrupt, and a zonation or physiographic succession of communities can be recognized. This transition from the ocean to fresh water and from the ocean to land as we see it today is of special interest since it parallels the probable evolution and dispersal of life in past ages.

Life is generally believed to have originated in the littoral region. Apparently no great group (phyla) of animals originated except in the ocean. The routes by which animals probably left the ocean and reached fresh-water and land have been various. Some animals probably migrated directly across sea branches; other probably ascended rivers, passed through marshes and swamps, or burrowed through soil. Some animals were transferred from the ocean by land elevations which isolated them in bodies of water which gradually became fresh Imigration from the sea did not take place at one time. It has occurred many times in the past and is slowly progressing on many shores today. The most successful animal colnizers of the land have been: (1) the arthropods, which have in many cases developed booklungs or tracheae for breathing air; (2) The vertebrates, with lungs and dry skins; and (3) the snails, with slime and spirally coiled shells to prevent desiccation. There are at present many examples of animals which are in the midst of their transformation form marine to fresh-water animals, or from marine or fresh-water into land animlas. Not only have plants and animals emigrated from sea to land, but there are countless instances when migrations have taken and are taking place in the opposite direction. Grasses, insects, reptiles, birds and mammals have left the land for the sea.. Fishes began in fresh water, but now range through the ocean at all depths.

On rocky shores and cliffs there is a splash or suprallittoral zone above high tide level. Green algae occur here and scattered individuals of marine sanils, acorn barnacles, limpets, amphipod sandfleas and flatworms, as well as insects, especially. Diptera and other forms that come from the land. Above the influence of splashing, the rocks may be covered with lichens and mosses, representing the initial stage's in the terrestrial rock sere. However, salt spray is often blown inland a considerable distance to affect conspicuously the development of normal terrestrial vegetation and its acompanying animal life. Cliffs along the ocean, as well as sandy beaches and islands, are favorite nesting places for large numbers of pelagic birds.

3 Energy from the Sea

Sea Currents

The water masses in the oceans are not dormant bodies but are in constant motion. If the movements of the water are on established lines they are referred to as marine currents. The causes of these currents are very different. The most important factors affecting marine currents are wind, density variations within the water masses, and differences in the heat flow between water and atmosphere.

When these factors occur in combination they may serve to reinforce one another as well as to neutralize one another. For the most part of the direction of individual ocean currents is determined by the earth's rotation the shape of the continents, submarine ridges, and tidal currents. Owing to the multiplicity of determining factors, any estimate of the direction of the resulting marine current is extraordinarily complicated. The velocity of the ocean currents compared with those in rivers is substantially lower.

If the wind sets the surface water in motion, the direction of the current does not necessarily follow that of the wind. Owing to the rotation of the earth and the Coriolis forces generated thereby, the currents in a particular sector are linked to the direction of the wind. In the northern hemisphere all the currents are deflected clock wise to the right, and in the southern hemisphere to the left. Since at the same time the prevailing winds are deflected likewise, the deflecting forces receive additional reinforcement. The deflection of the water currents is thus all the stronger the deeper the water columns are below the surface.

Basically, two types of marine currents are distinguished: surface currents and underwater currents. The most important surface currents are found in the equator belt between the northern and southern tropic zones. The northest and southest trade winds provide the motive force for these big equatorial currents. Between the individual continents these currents are diverted in their directions and deflected to the north and south. Since the Coriolis forces are intensified as the geographical latitude increases, the original current direction is gradually reversed.

Consequently a powerful flow of surface currents develops. In the middle of the cycle, zones are encountered with relatively modest surface movements (for example, the Sargasso Sea in the North Atlantic). The pattern of surface currents is roughly the same in all the oceans. The water masses that flow north and south from the equator toward the poles cool down gradually on the way, more and more rapidly as the geographical latitude increases. Their density rises in the process, so that water masses sink to a lower level as they go and eventually return to the equator as cold seabed currents. In all cases in which the wind carries off big masses of water, these must be replaced by a return flow. This happens as a result of either an influx of surface water or an influx of water from a lower level.

Water masses flowing toward the poles and so becoming cooler set off the deep water currents. Regions in which surface currents combine to produce deep water currents are known as convergence areas. Under the influence of surface, deep water and sea bed currents, a constant exchange of water masses develops in the course of time in all the big oceans.

If it were possible to utilize a mere 0.5% of the head stored in the form of temperature differentials in the oceans of the world, man's energy problems would all be solved. This concept, however, utopian it may sound, is not in fact utterly realistic. Sea-heat power plants fall within the field of technical possibility. The idea of utilizing the energy stored in the ocean is not new. It was first suggested in 1881 by the French physicist Jacques d'Arsonval. His fellow countryman George Claude erected the first sea heat power plant in Cuba in 1929. Its power output then amounted to 22 kilowatts. Its performance was satisfactory, but it was uneconomical, as Claude's technical resources were inadequate.

American environmental engineers have now developed the idea. Design work has begun at the University of Massachusetts for a sea heat power plant to be erected in the gulf stream of the coast of south Florida. The theory of a sea heat power plant is as follows: between various water depths there is normally a temperature differential which can be utilized for energy production.

In the tropics, that is between the celestial spheres of Cancer and Capricorn, the surface waters of the oceans have an almost unvarying temperature of 25°C (77°F) over the whole year. At a depth of 1000 meters (3200 feet) or so, on the other hand, the prevailing temperature is only 5°C (41°F). This temperature differential can be utilized as in conventional power plants by letting the liquid on the "warm side" evaporate or simmer. The steam drives a turbine and is condensed on the "cold side."

As the heat differentials are in a narrow range, water is not suitable as a medium. First calculations indicate that propane gas would be a very suitable gas for circulation in sea water power plants. It boils at temperatures of under 25°C, and can return to liquid form at 5°C. The power plant could be installed on a floating island in the open sea.

Power plants of this kind could provide a realistic technical solution for the utilization of solar energy, for the water is heated by the sun, and the oceans serve as energy stores. The temperature differentials arise as a result of circulation between the poles and the tropics. Sea heat power plants could also acquire much greater importance.

The cold water pumped up from the depths of the sea could serve to do more than jut cool the turbines. It still remains cool enough to condense water vapour drawn off from seawater into a vacuum chamber. In this way the power plant can extract drinking water from the sea. It is estimated that a plant with an output of 100 MW of electric power could provide some 200 million liters of drinking water a day. In contrast to the development of nuclear energy, especially by fast breeder reactors, no serious new technical problems are posed by this form of utilization of solar energy.

The periodic alternation of ebb and flow on the sea coast is caused by forces that, in addition to wind and the earth's rotation, are exerted by the influence of sun and moon on the water masses of the oceans. The flow of the tides is highly complicated, and varies from place to place. Since the days of Isaac Newton the attraction forces of the sun and moon have been recognized as the cause of the movement of the tides. Their influence is vastly different in strength, however, and the stronger effect must be attributed to the moon.

Two separate lunar forces have to be distinguished; the attraction power of the moon and the centrifugal force arising from the rotation of the moon and the earth (at the same point of concentration). Whereas the centrifugal force remains constant at each point of the earth's surface, the attraction force of the moon changes in inverse ratio to the square of its distance from the earth.

The attraction force of the moon on the side of the earth facing away from the moon is less than that on the side of the earth facing it. The result of this is a horizontal force that tends to concentrate the water masses of the earth at two specific points, of which one is directly facing the moon and the other is on the precisely opposite side of the earth. At these two points (A and B in the diagram) tidal crests are thus produced. In consequence of the earth's rotation and of lunation (the approximately 29 day period between two successive new moons) in an ideal situation, these would travel round the earth in a period exactly corresponding to the length of a lunar day (24 hours 50 minutes).

The sun two exerts tidal forces on the earth. The sun's mass is, of course, substantially greater than that of the moon, but it is also at a very much greater distance from the earth. Because of this, the tidal forces of the sun amount to only 46% of those of the moon. The change of tidal forces in the course of a month emerges from the interplay of the forces of the sun and moon. Thus, the pull from the sun and the moon reach their peak at times of full and new moon and combine to produce spring tides. In the first and last quarters of the moon the orthogonal directions of their forces are directly opposed to one another, and so lead to a reduction of the tidal forces and result in the occurrence of neap tides. Spring tides and neap tides occur twice in the course of a lunar month.

The position of the continents prevents the tides from registering ideal periodic changes and so produces many irregularities. As the vast water masses, once they are in motion, record a natural period of oscillation, the periodic changes in the tides and the natural motion of the water can overlap and so produce very complicated tidal phenomena. On occasion this can result in resonances, which will produce a maximum expansion of the tidal range (upto 14 meters), as can some times happen on the coast of Brittany. In tidal power plants an attempt is made to use this energy for human needs. Apart from the tidal forces, wind and air pressure have considerable influence on the extent of the tidal variations. Thus constant wind pressures are capable of producing independent tidal ranges of about 1 meter. Storm conditions sometimes give rise to extreme anomalies which manifest themselves as storm tides.

Tidal Power Plants

Tidal power plants are a particularly attractive way of utilizing the regenerative forces of nature for purposes of power production. Solar energy and lunar energy can be converted into electric power by this means in the same way as in a water works. The process extends over a period of 12 hours and 24 minutes, that is, half the period it takes the moon to circle the earth.

As the moon is the nearest of all the heavenly bodies to the earth, the force of gravity exerted by the moon has a measurable influence on the earth. In the big water zones of the earth (the oceans) the result is that where the moon is directly overhead, owing to its attractive power, the water level is raised somewhat (flood tide), whereas in other regions it will be below the norm (ebb tide). These two phases are particularly sharply delineated on the coastlines of seas and oceans.

At specially favourable points on the coast, tidal differences can be distinguished of 10 meters (32 feet) and more. If at such a point on the coast, a bay or a river mouth is shut off by a sluice of floodgate, energy can be obtained with the aid of turbines driven by water currents set up by the alternate filling and emptying of the bay.

The maximum power produced by a tidal power plant is determined by the potential energy of the water mass stored in a basin during the flood tide and by the tidal period. The coefficient of efficiency of this energy utilization lies between 20% and 25%. The pioneer in the erection of tidal power plants was West Germany. In the First World War a small tidal power station was in operation in Husum, using the flow and ebb of the current, in a former oyster-culture basin, in order to supply houses in the neighbourhood with electricity.

The first big tidal power station came in to operation in 1966 at the mouth of the river Rance in France. Today its output amount to 240 megawatts, and should be raised eventually to 320 MW. Altogether, 10 turbines are set in a dam 750 meters long, and they work in both directions, using both the incoming and outgoing tides. The tidal range (the difference between high and low tide) averages about 10 meters, with a maximum of 13 meters. In 1968 a small plant of 400 kilowatts was started up in the USSR at the mouth of the Kislaya, 80 km northeast of Murmansk. A larger plant of 320 MW is projected for the Lombowska River on the northeast coast of the Kola Peninsula.

The following table shows part of the power available on the earth from tidal energy. Possible performance in the various locations ranges from 2 to 20,000 MW. The total potential performance for all these locations amounts, according to preliminary estimates, to

64,000 MW (compared with the output of a present day nuclear power plant of 1200 MW).

Table 1
Operational of a tidal powre station when the tide begins to EBB

Place or region	*Average potential output (MW)*	*Possible annual Power production (MWh millions)*
North America		
Bay of Fundy (9 locations)	29,000	255
South America		
Argentina (San Jose)	6,000	52
Europe		
England (R Severn)	2,000	15
France (9 locations)	11,000	98
USSR (4 locations)	16,000	140

Glacier Power Plants

After many years research work in Arctic stations, where he was conversant with the structure and dynamics of the Greenland ice sheet, a Swiss hydrobiologist came up a short time ago with an unusual project for energy production. Enormous quantities of ice melt in Greenland every summer and run down into the sea from land 2000-3000 m (6500-9800 feet) lakh.

If normal power stations were built on the coast in Greenland, enormous quantities of electric power could be produced by this means. The technology of the necessary hydraulic power stations would be standard, but the construction material for the dams and the supply and pressure, ducts would be something quite new, namely rock hard glacier ice. As the biggest island in the world, Greenland has a surface area of about 2 million sq. km. (760,000 sq.m.). On a conservative estimate something like 500,000 sq. km. (190,000 sq. mi) would be available for the installation of storage basins.

With the sun shining for about 2000 hours per annum, some 8 kwh of energy could be produced each year from 1 sq. m. of storage lake area. From the estimated 500,000 sq. km of storage lake area, about 4000 billion kwh of useful energy should be obtainable each year. This represents an electric power output of about 200,000 MW, approximating the output of 200 large atomic power stations.

During the summer an extensive canal system, corresponding to a network of streams, would have to be built inside Greenland. The ice melted by the suns heat would run along these canals into big storage lakes. In the course of the flow of the melted ice water, which would be warm compared with the ice it came into contact with the streams would expand of their own accord, so that relatively little technical energy would be required for the construction of the canal system. Owing to the differences in elevation from 2000 to 3000 m, the melted ice water would shoot through ducts from the storage lake down to normal, big hydraulic power stations installed on the coast.

The storage lakes are required mainly to ensure that energy production will be available also in winter. In the winter a layer of ice 1 to 2meters thick forms on the surface of the storage lakes and protect the water below from further freezing. The water below the layer of surface ice could thus be drawn off for energy production as required. The necessary barrages, and the supply and pressure ducts, can be modeled by suitable techniques with the aid of the ice. Tested procedures already exist for the construction of ice channels in glaciers.

By means of a circular cutting guide (a steel tube with an interfacing rotating saw tooth rim), cylinders, can be cut out of the glacier ice having a diameter of about 10 m (33 feet). With a subsidiary circular saw appliance the ice cylinder cut out lengthways can be split up into semicircular portions and lifted out on to the canal bank by means of a rotating table turned at an angle of 180°. These segments can be stacked together quite easily to form dams, whereupon they rapidly freeze up again.

By means of such glacier power plants, cheap energy could be produced in large quantities without causing any pollution of the environment and without the consumption of law materials. For transmission of the energy there are two possibilities:

(1) The electric current produced can be conveyed directly by submarine cable to North-America and Europe; or

(2) This is probably the better solution, with the aid of the electric energy produced in the hydraulic power station the water can be split up into hydrogen and oxygen, and the hydrogen can be liquefied and conveyed by tanker to the country of consumption. It could then be used for heating, the driving advanced motor vehicles, or in fuel cells for electricity production.

Petroleum and natural gas are the most important sources of energy and are the raw materials most in demand by the chemical industry today. For the most part, petroleum comprises a mixture of several hundred or even thousand hydrocarbon constituents. In many cases these are the only constituents, while in others a large number of other organic compounds are present, such as oleic acid, asphaltene, gums and a variety of sulfur and nitrogen compounds.

The vast proliferation of hydrocarbons stems from the tetravalenc of carbon and its many possible compounds in the form of chains of different lengths with single, double and triple bonds, of rings of different bond types and sizes and of the most varied bonding with one another: and it also allows the formation of side chains. Gums and asphaltenes are high molecular weight organic compounds, which give to the oils their dark colour and increase their viscosity. The sulfur and nitrogen compounds are in the form of chains or rings, in which carbon atoms are replaced by sulfur or nitrogen atoms. The explanation of how petroleum is formed is one of the most difficult problems encountered in geological research.

Almost all the facts point to its organic development from decomposition products of plant and animal organisms that were deposited in the mud beds of waterways, were preserved there from immediate putrefaction by the lack of oxygen, and then formed rock bitumen. One problem regarding this explanation of the origin is that organisms are composed to a large extent of oxidized compounds whereas petroleum consists of reduced (oxygen-free) compounds. Reduction is conceivable, however, only as a result of energy input. In the first phase of transformation at least, the necessary energy could not have been derived from the heat of the earth, as the petroleum would then still contain elements from the chlorophyll and blood pigment, which will not sustain a higher temperature than 200C° (424°F) and will persist only in an oxygen free environment. One possible reaction, which also occurs in lower temperatures and is anaerobic (without free oxygen) is biological reduction by bacteria. In the sediment in water and in oil a whole series of anaerobic bacteria can be recognized that are able to reduce organic substances and to transform them into fatty acids, methane, and ethane, and also into other hydrocarbons. The hydrocarbons in question do not form petroleum, however. The further transformation of the long chain compounds (fatty acids) into petroleum with its light paraffins (5-7 carbon atoms) can occur only as lower temperatures and in the absence of active clays as catalyzers or at great depths under high pressures by thermal decomposition. Under this thermal decomposition or fission, long chain hydrocarbons will be split up into methane and other light gaseous hydrocarbons.

Natural gas is produced in the same way, but contains carbon dioxide, nitrogen, hydrogen sulfide, and mixtures of rare gases in addition to hydrocarbons. A distinction is made here between dry gases, which consist almost exclusively of methane, wet gases (ethane, propane, butane, etc.), and acid gases with mixtures of hydrogen sulfide. The formation of petroleum and natural gas thus possibly represents a process that embraces some very different reactions (bacterial and thermal decomposition or fission) and extends over long periods. The formation of the most recent petroleum fields dates back roughly one million years.

Salt Deposits

In the weathering of volcanic rocks the compounds dissolved are mainly those of the alkaline earth metals calcium and magnesium, together with those of the alkali metals sodium and potassium, whereas the oxides of silicon are in effect insoluble. Part of the potassium is removed from the weathering solution by clay minerals, in which the potassium is incorporated. The potassium, calcium and sodium, along with magnesium, dissolved carbon dioxide, and sulfur compounds, are carried down the rivers into the oceans, where they eventually form salts.

In open sea water the salt concentration is too low for the salts to be deposited. Owing to evaporation of the water, however, the concentration can be raised to the limit, which results in the deposit of the precipitated salt. This presupposes the delimitation of sea basins, in which certain conditions must be present. The loss by evaporation must exceed the inflow of water. And the shores of the inland lake or of the sea bay must be so flat that no more material is fed into the basin from the land.

The first substances to settle are the least soluble carbonates and sulfates. This can also occur in very small concentrations biogenetically in the case of marine organisms which by their metabolism absorb the lime as a stroma supporting tissue substance in their scales. Some 40% of the sea bed is covered with what is known as foraminiferal ooze, formed from the shells of dead protozoa and from clay substances. In the further process of deposit and sedimentation, rock salt is produced, and finally the most easily soluble potash salt. The sinking of the bed of the basin, a rise in its shoreline and an inflow of new water masses will mean that the water formation process is repeated. The salt formation thus occurs only during a static phase in part of the earth crust. In many regions several such cycles can occur in succession. The clays washed into the basin at times of increased erosion of the surrounding land surfaces protect the salt deposits from being washed away.

Salt formation is possible both in inland waters and in the sea. The salts differ from one another, however, in their chemical composition, which is more astonishing in that not only the inland waters but also the seas receive their mineral constituents from the land surface. This involves the supposition that the oceans have an original supply of salt which is only slightly modified by the fresh supplies washed in from the land. As the various salts are deposited one after the other, they tend to remains in layers one upon another. Clay and carboniferous rocks are deposited first from the shore of the basin into the middle; sulfates and chlorides are deposited in the basin's interior. In this way salt basins are formed with their typical salt distributions.

From salt deposits, which are overlaid by sediment, salt masses can form. This depends on the fact that salt begins to flow even when there is very little difference in tension in the sediment. Differences in tension are produced, for instance, by movements in the earth's crust or differences in erosion levels in the overlying rocks. The salt then flows into zones of lower pressure, and accumulates to form salt cusions. At this point the overlaying rocks are raised; and at the point at which the salt is carried away they subside.

4 Marine Resources

It is estimated that an area of 10 million sq. km. in the Indian Ocean, both inside and outside the respective exclusive economic zones, is covered by 150 million tonnes of polymetallic nodules, next only to the Pacific Ocean. By 1992, India could start commercial production of nearly 3 million tonnes of dry nodules a year to replace and reduce its import of nickel, copper, cobalt and manganese. Besides, India has world's one fifth marine area with rich fauna. However till now the major fishing area is confined to an offshore belt of 1 sq. km. only, and the rest of the Exclusive Economic Zonel (EEZ) is untouched, which abounds in fishes of lage variety. Also, the 1981 estimate of India's offshore oil reserves were around 1100 million tonnes and the gas reserves being 12,000 ´ 10^9 cubic feet. But even then she is notable to meet her 50% of domestic requirement. Lack of coherent maritime policy, unplanned ocean management system, inherent technological backwardness, in addition to her comparatively new awakening towards ocean resources and maritime trade, are perhaps the main causes for her being underdeveloped in this field. A brief survey of the EEZ resources of India can be categorized under two categories namely living

(1) nonliving,

(2) living.

Nonliving Resources of India's EEZ: A General Account Growing attention towards exploration of oil and gas, particularly after 1973 world oil crisis, mostly in the continental shelf region, and a distinct shift towards mineral exploration and exploitation in whole of the EEZ after the final declaration of UNCLOS III in 1982, other than regular search for non-conventional sources of energy within the said region represent the overall marine activities taking place in the fields of scientific and economic development taking place in respective EEZ.

The main reason behind the enhancing research programs by both developing and developed nation is the increasing demand of mineral requirements by modern societies, in an almost undefined and unpredictable volume, supplemented with the fact that land resources are either on the verge of depletion level or otherwise their vital importance has been realized as a reserve in near future.

Found mostly beyond EEZ, manganese nodules and polymetallic sulphides are of special value, as they consist of strategic metals, viz., cobalt, platinum, chromium etc., in a concentrated form.

Classification of nonliving resources have been made on various basis. In the continental margin area unconsolidated sedimentary deposits include sand and gravel, calcium shells, placers, beaded phosphorites, metalliferous or brine rich sediments, while barities and phosphorites form the major modular and crust deposits.

Selective sorting of mineral grains occurs by mechanical winnowing of wind, waves and currents, and by gravity setting and concentration. Among the selected production figures for sand and gravel from marine environment, UK and Japan top the list by producing more then 165,000,000 tons, having value more than \$330,000,000 (USD) and 84,700,000 tons having value \$170,000,000 (USD) respectively, mostly from a depth ranging between 35m-50m.

Calcareous sands, shells and coral sands are of course renewable biological resources, but they need special care while exploited at a commercial scale, because of their direct effects on the coastal marine environment.

Even offshore bank exploitation may have an impact on the beach amenities. Found in nearshore zones, placer deposits are mechanically concentrated heavy detrital materials which originate from eroded offshore rocks, often discharged by river work in considerable amount, in the form of black or dark stringers and lenses on modern beaches.

Past sea level changes, recurrent wave actions and dominant tidal currents have however produced major locational changes, including their nature and depth of occurrence. While buried placers are little known, cassiterite exploration in Southeast Asian tin belts undoubtedly will reveal several examples, although this still remains a fact that general placers are richer on modern beaches than on submerged ones. Further wider shelves have greater number of placer deposits.

A comprehensive account of the United States' EEZ placer deposits was recently given by Beauchamp and Cruickshank including its comprehensive economic futurology. In India, National Institute of Oceanography has taken the sole responsibility of mapping, exploring and exploiting these huge deposits, mostly available on the western coast of India, in the form of tin placers, complex heavy minerals, precious metal placers and rare diamond accumulations.

Besides biotite, biotite-muscovite, and cassiterite, the tin placer deposits contain pyrite, ilmenite, zircon, tourmaline and monazite, commonly found in India, Burma, Thailand, Malaysia, Indonesia, Cornwall (UK) etc.

Among the complex heavy mineral sands containing principally ilmenite, rutile, zircon, monazite, garnet, and magnetite, the iron-titaniferous sands have been identified at commercially exploitable level at Southern Japan, East and West Australia, California, Oregon, Alaska, New Zealand, India, Bangladesh and Sri Lanka mainly, besides USSR, Poland, Mozambique, Italy, Philippines, Argentina, Mexico and Brazil.

As regards monazite, mineral placers were identified especially offshore India and Sri Lanka. Precious metal placers containing gold, platinum and related metals have been mainly found concentrated in Alaska, Oregon, California, Nova Scotia, Lapter Sea (USSR), Southeast Australia, Bering Sea etc. Diamond accumulations containing mainly kimberlite, epidote, ilmenite, magnetite, and pyrite has its only known commercially exploitable deposit at Hottentot Bay near Orange River in Namibia.

Beaded phosphorite minerals mainly in the form of carbonate fluorapatite are found in the upwelling areas like Peru-Chile slope, North and Southwest Africa, Western Australia, and in the continental shelf areas of Congo, Baja California. Metalliferous and brine-rich sediments, known in sufficient concentration in Red Sea Spreading Center, contains mineral assemblages like clays, hydrous oxides, carbonate and sulphides.

However the most important brine deposits are found in the Atlantis II Deep, where economic estimates suggest about 3 ´ 10^6 tons of metals, excluding iron which forms almost 29% of the deposits. Nodule and crust deposits have been reported present, although in less exploitable quantity off southern India and in the East Indies region. Phosphorite nodules and crusts, generally present on submerged marine plateaus and continental rises viz. Blake Plateau, the California Banks, Agulhas Bank, the Chatham Rise, and in some scattered seamounts.

Known manganese nodules deposits of both mononucleate and polynucleate types are most abundant in deep seas of Pacific (in Equatorial North Pacific nodule belt, Southwestern Pacific Basin, Peru Basin, Chile Basin, South Tasman Basin and Northern Tahiti Basin), Indian Ocean (South Equatorial basin, Crozet Basin, Central Indian Basin and Western Australian Basin) and in Atlantic Ocean (in Western Atlantic Basin, Drake Passage, Western South Atlantic and Blake Plateau).

Much interest is presently focussed on occurrences of cobalt rich ferromanganese crusts on seamounts and oceanic crusts, some of which have already been discovered within the EEZ of USA. Massive deposits of polymetallic sulphides have attracted recent attention. They are generally found at mid-oceanic rideges viz. stratiform sulphide bodies at Atlantis II Deep, Red Sea, sulphide mounds surmounted by chimneys at Eastern Pacific deposits and coalesced mounds surmounted by chimneys at Galapagos Spreading Center with much concentrated quantities of zinc (Zn.), copper (Cu), and gold (Au) besides some traces of arsenic and cadmium.

Although exploration in Pacific have yielded much regarding their location and quantity, the possible existence in Indian Ocean, that too in considerably exploitable amount, is yet to be tested and quantified in economic terms. In India, much emphasis has been there during the Sixth and the Seventh Five Year Plan regarding the surveys to locate and map the availability of living and nonliving resources of the Exclusive Economic Zone and that of the continental shelf. Different governmental agencies viz. Department of Ocean Development (DOD), National Institute of Oceanography (NIO), Oil and Natural Gas Commission (ONGC), Oil India Limited (OIL) etc., with their maximum emphasis on fisheries and hydrocarbon resources, are burdened with solving the problem of energy requirement of this world's second most populated country.

Further, the northern Indian Ocean is one of the few oceanic regions of the world which comes under the influence of the monsoon. The uniqueness of this region, therefore, lies in the seasonal and interannual variation that it exhibits. Little is known about the nature of this variability and lesser still, about the causes behind it. Rigorous efforts are being made by sophisticated research vessels like ORV Sagar Kanya and FORV Sagar Sampada.

Their utilization for some years has been going up to 82.6%. Besides, other individual and collective efforts by scientists all over the country is directed in a multidisciplinary format. The survey of the entire EEZ, according to DOD, is expected to continue for a decade or so. Intensive surveys are being undertaken to map out the nonliving resources especially minerals and hydrocarbons. Estimation of potential fisheries resources and its maximum sustainable yield is being made.

The important considerations which need immediate attention from different governmental and nongovernmental agencies regarding India's EEZ are:

1. Clarification of the level up to which present EEZ utilization is going on;
2. Demarcation of optimum limit of EEZ utilization both in terms of exploration and exploitation;

3. Projected need of marine substances both living and nonliving from the EEZ by 2000 A.D., by a realistic evaluation of both protein requirements and energy substitution in different sectors;
4. Technological requirement for the fullest exploitation of EEZ resources and extent of foreign collaboration or technology transfer to be required by 2000 A.d. for its substainable exploitation;
5. Fixation of priority level, to be given to the frontier areas of ocean development in India and share of EEZ management in that, particularly in the fields of food (protein), minerals, transportation, recreation and energy, all aiming at improving India's socio-economic conditions;
6. Planning of offshore construction and required operational structure in total and the investment to be incurred.

Nonliving Resources of India's EEZ

(A) *Freshwater from sea :* The desalination technologies under consideration by the concerned governmental agencies are (1) by solar stills, (2) by flash distillation, (3) electrodialysis, (4) rever se osmosis, the last being most widely used, generating capacity of plant generally varying anything between 10,000 to 15,000 litres per day, most useful for the dry coastal areas of Indian Peninsula, presently drinking brackish water with salinity ranging from 5 to 20% in most of its rural areas.

Central Salt and Marine Chemicals Research Institute, Bhavnagar has already done necessary research and development work and has made available the technology of "reverse osmosis" to BHEL, for setting up desalinization plants. One such pilot plant has been recently installed at Puthagram, near Madras. Work on the second plant with same capacity of 50,000 liters potable water per day is continuing and is expected to be installed in the East Godavari district. In addition of salt, the system also remove harmful matters, like bacteria from brackish water. Besides it saves much energy as compared to the conventional thermal desalination system, thus being economically viable alternative.

Although a serious threat to future large scale desalination program at global level, particularly for the vast desert areas along the oceans, is from the comparatively cheaper huge icebergs to be fetched from southern oceans, although ita waits further technological advancement and a major breakthrough in the understanding of the mysterious characteristics of icebergs, particularly the dangers to be encountered by its cleavage in any phase of its displacements. However, for localized use in coastal areas, desalination process is still economical and rather indispensable.

(B) *Drugs from the sea:* Although this utilization of marine resource is of much recent origin, by now at least 100 plants and small marine organisms have been recognized for the manufacture of important lifesaving drugs in India. More than forty are of special value, some of them containing remarkable antifertility properties, viz. Glacilaria species, with sufficient prostaglandins. At present a great need is there in India to set separate shell for studying and culturing these marine microorganisms, both of floral and faunal nature, and to help successfully in deriving and synthesizing the bioactive substances which are useful in curing many incurable diseases.

Studies in India clearly suggest that all the species of marine algae, those exhibit antifertility properties may also contain prostaglandins. Their multiple use and suitability to human health is being further researched, but the most important aspect still lies with the fact that their potential presence either in the form of deposition or dissolution is still not economically beneficial for large scale economic exploitation with the existing state of technology, understanding of distribution pattern and density.

(C) *Marine chemicals:* Lesser quantitative concentration and difficult extraction methods are the major hurdles in deriving most of the chemicals economically, about 60, from the seawater. Commercial exploitation as yet has been successfully managed only for six chemicals viz. sodium, chlorine, magnesium, bromine, calcium and sulphur.

Greater possibility is also there in future for deriving potassium, iodine, uranium and gold from seawater. However, optimistic proposals supported by major research findings in this field leaves us with the only hope that by 2000 A.D.

India will be able to make use of its EEZ waters, mainly coastal, in deriving at least the rare chemicals which are dissolved in seawater and the same time are not easily available on land, provided concerted effort is being made to develop sufficient technological capability to meet the requirements of the complex stages involved in this process.

Besides some marine organisms concentrate trace elements in their bodies by a factor of 10,000 to 1,000,000. Cultivation and harvest of such organisms could perhaps provide a way of concentrating trace metals dissolved in seawater so that they can be commercially recoverable by biochemical means.

Seventy-one per cent of the total solar energy i.e. about 106 watts reaching the earth is absorbed by oceans, thus blessing the later with the huge/energy potential. Apart from being renewable and continues the added attraction of ocean energy forms lies in their being pollution free, resulting in remarkable minimization of health hazards, and the great competence for fulfilling the local energy requirements particularly offshore. The leading countries in this field of research are USA, Japan, France, UK and the Netherlands, implementing mostly the total systems approach, successfully overcoming the attached disadvantages.

Costal inlets are best for tidal energy resources. Technology for the utilization of tidal powers in all possible physical situations is already being developed. La Rance, since 1967, has been successfully supplying 240 MW of power to the French electric grid. In the same way, nearshore tropical and subtropical oceans are best for vertical temperature gradient, thus for Ocean Thermal Energy Conversion (OTEC).

Areas of strong persistent currents, such as most of the western coastal boundaries of the world oceans are almost suitable for extraction of energy from ocean currents. River outlets are most suitable for extraction of -the salinity gradient energy. Similarly coastal areas with large persistent waves can contribute maximum wave energy while wind turbines located in artificial or natural islands, near the shore, can generate maximum wind energy.

In India, research and development activities are going on a large scale regarding the derivation of above mentioned forms of ocean energy. However the forms of energy which has already been derived or could , be derived in the very near future with sufficient technological feasibility are OTEC, wave energy, and tidal energy only.

(D) *OTEC:* OTEC utilizes the temperature difference which exists between the surface seawater (28-30°C) and the water occurring at a depth of 800- 1000 metres (5-7°C)

: to generate power through the medium of a suitable fluid which vaporizes at high temperature and condenses at low temperature and in this process turns a special type of turbine.

According to feasibility studies OTEC power is economically competitive with fossil and nuclear powers. Directives establish a US national goal of 10,000 MW on line by 1999, from OTEC facilities only. Some of the best potential sites are located near the islands of Lakshwadweep and Andaman and Nicobar Islands groups. Only two islands, namely Kavaratti and Minicoy were found to offer possibilities for installing on OTEC plant, preferably each of I MW potential.

From the available bathymetry and temperature profiles around the islands, it has been found that a depth of 1000 metres will be available within the distance of about 3 kin from the islands and a temperature difference of 20°C between the surface and 1000 in depth throughout the year.

Therefore, it has been considered that a shore based plant is ideally suited for these island sites, which also have big lagoons on western side. These islands are of a coral origin and detailed studies for placing the cold water pipeline on the sea-shope are necessary. Kavaratti was found relatively more uniform.

Besides, in the technological field, procurement of metal components viz. plate heat exchangers, cold water pipes, turbogenerators, pumps, etc. are most vital for successful exploitation of OTEC potentials. Technology to fight corrosion and bio-fouling, for longer life of installed offshore machines, is also of vital relevance.

The law of the sea is silent on the rights and obligations derived from energy extraction from the high seas and on disputes arising from such extraction, for instance, by a grazer, within a country's EEZ, upstream from another nation's EEZ, and on the issue discharging its return water above the thermocline, thereby degrading the downstream resources.

While the worldwide OTEC potential is expected to be around 10 million MW, the same around the Indian coast has been estimated to be 50,000 MW (i.e., 0.2%) only. Around 3000 kin along the Indian coast down from Bombay on the west coast up to Visakhapatnam on the east coast, a temperature difference of 20°C throughout the year is a available. Thus India has more than 3 lakh sq. km of tropical waters in her EEZ where sufficient temperature gradient exists throughout the year.

Thus while estimated global OTEC potential is approximately 10 times the total world electric power production today, in India, the OTEC potential is about 15% of our total installed power generating capacity, hence necessitating the indigenous technological development and rapid scientific transformation relevant to this field.

According to one estimate by Kurachi, approximately 1 million tons of cold water must be pumped up per hour to produce 100,000 KW of electricity from OTEC plant 13, i.e. power needed for pumping operations alone is as much as 10% of the generated power.

Even then, OTEC seems to be a useful addition to energy arsenal of the present world, if anyhow brought to the level of larger cost effectiveness. But it is too early in the development process to make any kind of economic assessment of the potential role of OTEC plants in the national and international energy picture. For what they are worth, cost estimates for very small plants seem very high even at experimental level.

But what this means in terms of a possible full 400-500 14 megawatt plant, is simply impossible to determine at the moment. The history of technology transfer from government to prime industry and from pilot plant to operating levels is not one that would generate optimism.

The USA with ultimate plan of 400 MW plants at and around isolated islands near Hawaii, Japan with 10 MW floating plants and 20 MW land based plants on the island of Nauru, France with 5 MW land based plant on the island of Tahiti, the Netherlands for islands of Ball, Sweden with proposals for the islands of Jamaica, UK with 10 MW floating plants and with plans of commercially profitable 40 MW plants, have either started functioning or are scheduled for the same in near future. The sums of money are very large, may be around US$60 million for a 10 MW floating plant.

However in the Indian situation, as elsewhere, this baseload OTEC resource, by both of open and hybrid close cyles, seems to be of practical value favourably to southward situated islands where modest contribution to small and medium scale intraregional energy needs could be met successfully.

(E) *Wave energy:* Global wave energy potential is of the order of 45×10^{18} Watt. Variation of annual average value can be, for example, from 3 KW/m of coastline in the Mediterranean to 90 KW/m in the North Atlantic. The seasonal variation is remarkable i.e. in winter 6 times greater than the summer one. The principal recent studies have been made in UK, Japan, but interest has also been shown by France, Norway, Sweden, India, and the USA.

Inventors have proposed many devices for large-scale wave energy collection and transmission to land, including, oscillating water columns, floats, moving pistons, rockers or paddles, and deformable bodies of both mechanical and pneumatic types. However on a small scale, wave energy already provides the modest power needs of remote navigation buoys, including a large one, Kamei, in the Sea of Japan.

In general, a 100m dome with 11 in diameter and 20 in high dome could generate 4000 KW from 2 meter high waves, a system known as dam-atoll system, made firstly by Lockheed Corporation. In simultaneously acts as a breakwater also which in turn, would have effects on the littoral movement and disposition of sediments.

Thus, apart from small impediment to navigation and commercial fishing, the major problem of wave-power recovery has been from the very hostile marine environments viz. strong storms and winds. Besides, it may also be feasible to combine power dissipation with power extraction in future breakwater construction.

The use of wave power by navigation buoys will continue and should expand because it is a good example of on-site energy use, particularly at islands. However, at this primary stage of scientific achievement in this area, estimates of capital costs are highly speculative and subject to greater uncertainty, particularly when taking into account the likely future increase in the costs of energy derived from nonrenewable and depleting fossil fuels.

In India, available informations based on visual observations indicate that annual average wave height along the Indian coast is of the order of 1.5 in with the wave period of about 6 seconds. But scientific data collected off Kalpakkam and near Kakinada on east coast, using wave rider buoys, indicate wave heights of more than 2.5 to 3 m, even during non-monsoon period.

This suggests that the annual average wave power potential along the Indian coast may lie anywhere between 5 KW/m and 15 KW/m. For total Indian coast, the wave energy potential is around 6,000 MW.

Multipurpose Wave Regulatory Systems (WRS) are being employed and fixed near the appropriate sites off the Indian coast, so that further designing of wave energy devices could be done after taking into account the collection of climatic data, prospects of coastal protection from wave erosion and about other uses of coastal waters viz. aquaculture, transport, harbor developments, etc.

Indian wave regulatory systems have been proposed to be set off the Madras coast at approximate water depth of about 10 meters and a distance of about 500 meters from the shore. Thus in India, a wave energy system exclusively for generation of electricity, does not seem to be economic.

A multifunctional system is being employed after incorporating the modifications to the oscillating water column reported in 1984 in Norway (i.e. The Harbour Concept). One twin turbine motor of 150 mm diameter has been successfully fabricated for demonstration, besides completion of studies on variation of wave energy at different depths, thus finding the optimum water depth, wave refraction studies in different lengths, completion of demonstration model on wave energy including a wave maker for a 1.2m wide flume and an oscillating water column to absorb the waves with a matching turbine proven in the 4 m wave flumes and generator to produce electricity.

(F) *Tidal energy.* Today France, the USSR, the UK and China are in the forefront of tidal energy development while the USA, Canada, the Republic of Korea, India and

Argentina having made serious efforts towards its advancement, both in single and multiple type basins as well as single tide or double tide type.

The peak efficiency of tidal power plants can reach 80%-90%, but for the simplest systems, the baseload is nil. Of the tidal power schemes now operating, only one is commercial in scale since 1967 with an output of 240 MW at LaRance in France (with coast approximately US $120 million and produces electricity at a cost of 3.5 cents/kWh.

The second one is an experimental one at Kislaya Cuba on Russia's Kola Peninsula in Barents Sea with some 400 kW capacity. In China large number of small plants in function is reported, while two sites of extreme potential are being developed by Canada and UK in Bay of Fundy and in Severn Estuary respectively, the later with estimated supply of 10% of the total England's annual demand for electricity.

An alternative location across the Bristol Channel, west of Minehead, would even yield twice this amount. A recent review of the economics of the proposed tidal energy schemes in the bay of Fundy concluded that they were equivalent to the nuclear plants and markedly superior to the coal fired thermal plants.

The total tidal resource of world oceans has been estimated at 3 x 101 megawatts. Only 2% of these, at relatively 30 sites around the world, is suitable for harnessing at economically feasible level. Even then, it is significant because it represents 5% of the 1980 worldwide power generation from all sources.

In other words, it is estimated that tidal power could displace between 5% and 10% of other energy sources in the area. Further, the plants have high reliability and a long life of around 75- 100 years and appear to increase four times in numbers in coming 50 years.India's Gulf to Kutch is one of the most suitable location. Work has been in full swing to complete the festibility report by March 1987. The other possible sites, according to 1966 survey, are the Gulf of Cambay and the Sunderbans, having annual average tidal range of 6 m and 3.5 m respectively. The maximum tidal range found, yet in India, is in the Gujarat offshore areas i.e. around 11 m.

According to Prof. E.M. Wilson, installed tidal capacities in the Gulf of Cambay, the Gulf of Kutch, and the Sunderbans are about 7300 MW, 100 MW, and 15 MW with estimated cost of around Rs. 1925 crores, Rs. 600 crores, and Rs. 15 crores respectively. The *7300* MW plant at Gulf of Cambay, if installed, will be able to contribute as much as 20% of India's total installed power capacity today. Realizing the potential, Gujarat State Electricity Board and Department of Energy, Government of India have jointly taken up a detailed project study in collaboration with Electricity de France, along with 9 other interested government agencies.

The project envisages construction of three main dams over the Hansthal Creek, the Sara Creek and the Phang Creek, having 3.1, 0.7 and 0.5 km. of length respectively. The three dams will be connected with each other by a 30 km. long dyke carrying a road on its top, however the only proposed powerhouse will be over the Hansthal Creek, functioning for 12 hours a day, 6 hours in each tidal cycle, producing 600 MW of electricity.

The CWPRS has already prepared a physical model of the tidal power project at Khadagwasla with a revised estimate of 5.3 crore, controlled by Kandla Port Trust. If the present siltation problem across here is solved anyhow, then this project will at least save Rs. 2-4 crores per year.

(G) *Currents:* Important currents, prevailing mostly in western parts of world oceans, include the Gulf Stream and the Guinea, the Brazil, the Mozambique, Japan (the Kuroshio) and the Australian currents. Both linear and rotary systems are being employed with large ocean coriollis turbines to derive this very diffuse energy source. For example, the Gulf Stream has the highest velocity (more than 2 m/s) and the power of its core with an average width of 50 km and depth of 120 in has been estimated at 24 GW.

Assuming utilization of 20% at an efficiency of 40%, the resulting 2 GW of electric power would not amount to a significant fraction of world's energy supply. Besides the efficiency amounts to little more than 50%. However, if suitable submerged generators could be designed, they might be placed in gaps between atolls, where currents are often violent.

Particularly if they are relatively close to the shore, these might be able to satisfy the limited energy needs of small neighbouring populations. Further, the navigational use of an ocean current can significantly reduce the transit time of ships travelling in its direction, which is a useful way of capitalizing on this energy source. The prospects of current energy on western coast of India is quite considerable, although not much development or technological breakthrough has yet taken in this field.

(H) *Offshore Wind Energy:* Placing wind turbines in offshore areas is environmentally advantageous. It reduces frictional losses. Although the net profit is only about 22%. Prospective site selection for installation of these turbines in India's EEZ, after surveying the different physical parameters in detail, is yet to be finalized. However, an imported wind generator has been installed for the first time in India to operate a lighthouse at Kanai Creek, Gujarat, with a power output of around *300* watts. The UK~ the USA, Denmark and Sweden are leading in this field of utilization of wind energy.

Offshore wind generator systems would have to be relatively large in order to defray the high cost of power connection to the shore. In real terms, a successful wind power system is the sail driven vessel. If only a small fraction of the world's commercial transport fleet could be replaced by wind assisted ships, a significant amount of oil could be saved, seemingly of much importance for Indian foreign exchange savings.

(I) *Salinity Gradient:* Extraction of 240 mb of existing equivalent pressure by use of semi-permeable membranes on a very large scale has the potential to provide an overall substantial amount of energy. Its development might capitalize firstly on its abundance and secondly on possible low efficiency small plant. Further the power density of salt domes and brines, which is much larger than normal seawater, might make it a favoured application for salinity gradient density.

World's major salinity gradient resource areas include coastal regions just offshore of major river basins such as Yangtze, Amazon, Congo, the Ganges an Mississippi rivers. In India, the Sunderbans and other large deltaic and estuarine points are potential places for generation of energy from salinity gradient.

Further i India, the total freshwater discharge from rivers an streams flowing into the sea amount to about 23 x 101 in sec., which could provide a power potential of about 54.8 101 MW. The energy density of this source is equivalent to about 250 in waterhead for a salinity difference of 35 ppt. However only a little work has been done on the technique of extraction of power from salinity gradient.

Other Sources

Although still at an experimental level, countries with larger EEZ have indeed possibility of extracting the following resources.

(J) *Magnetic ocean energy* production by magneto hydrodynamic generator (MHD) uses geotrophic currents and long strip that is wound with a flatwire, thus the flowing current produces electric current of 5 mho/m.

(K) *Marine biomass energy* in the form of CH^4 and CO^2 can be obtained mainly from Macrocystic and Glacilaria species. Current global estimates claim that marine biomass power could be competitive with conventional energy production. The prices would run about $18 per 100 m^3 with ayield of 112 tons/hectare/year. Major technological breakthrough is still awaited.

(L) *Marine chemosynthetic energy is* mostly associated with the vents near the seafloor great rifts, and from unusual marine faunas like worms, mussels, and clams found at depths around 2550 in and below using economically worthless H^2S for the purpose would be cheap and will also reduce pollution and eliminate a wate product, helping indeed in alleviating cost of the source material.

(M) *Geothermal energy* associated with rift zones and areas of intense volcanic activity are possible marine economic sources, particularly those who are coastal and insular in nature lying within the EEZ limit.

(N) *Heavy water viz.* deuterium and tritium, abundantly dissolved in oceanic water when fused at great temperatures releases great amount of heat, easily usable to generate electricity. Japan is seriously doing the pioneering work in its EEZ related to heavy water energy.

(O) *Radioactive energy* from marine sources is being developed. Japan Metal Mining Agency has started constructing experimental factories for the retrieval of uranium dissolved in seawater. Currently the Siemens Corporation of the USA is also attempting to use marine redioactive chemicals and radioactive wastes being dumped in oceans as energy sources. Current models are producing around 1.50 watts at a cost of $5 a watt and needs attention for possible economic viability after 5 years.

In a nutshell, we may comment in the field of other marine energy resources discussed just above, economic viability through capital investment and technological attainment, are the main points to be stressed upon in near future for any commercial level exploitation of these infinite renewable energy sources present in the ocean. The Indian situation in all these areas are yet at experimental exploratory level. Production, however, may be feasible within 15 years from now.

From the available geological and geophysical data, it has been estimated that the continental shelf around India up to a depth of 200 in, with an area of around 452,000 km^2, can have a potential oil reserve of 1 billion tonnes and natural gas around 271 bcm approximately. Recent offshore hydrocarbon discoveries indicate that the potential oil and natural gas reserve is very promising and we can look towards this in the coming 21st century with considerable optimism.

Most of the contemporary drilling activities in offshore area is carried out at depths of 200 in or so generally. Further, the Indian offshore oil reserves are of period Mioceane, Oligocene and Eocene.

Proven offshore reserves represent a quarter of the world total reserves. About 90% of these reserves are estimated to be under the continental shelves, mostly under the respective coastal states jurisdiction.

The overall strategy for the commercial development of offshore hydrocarbons depends on a careful appraisal of the potential for oil and gas and on technological and economic factors. Further much of the world's offshore oil and gas, mostly under EEZ, areas will be discovered and produced only if the relevant governments adopt a regulatory or fiscal policy appropriate to the realities involved.

Even the present reserve is considered to be maximum of 345 $\times$ 10^9 MT, feasible to be produced, it would not sustain consumption at current levels beyond the year 20 10. New findings of sufficient magnitude, able to sustain demand beyond the year 2000, would have to be made at twice the present discovery rate, but it can be visualized that even this optimistic scenario could only maintain A-situation in which oil and gas could be consumed as fast as it could be produced.

In India most of the known petroleum resources are confined to the narrow belt of tertiary strata which constitute the outer margin of the extra-peninsular mountains along their whole line of contact with the peninsular block of Deccan viz. the Rann of Kutch, the Cambay Gulf off Piram island (dome), Bombay High and the Andaman and Nicobar Islands.

The number of wells drilled in the offshore areas, by ONGC, was 36 in 1980 with a hole depth of 77,167 in against a total of 109 wells and 193,734 m of depth 32. According to the 1983-84 estimates, the ratio between production and consumption of oil in India is 1:1.3 approximately.

New discoveries have added to the attainment of self-sufficiency. Reduction in production as well as in consumption and efforts to supplement the fossil fuels by nonconventional and renewable sources of energy can only take us well within the 21st century, probably up to 2020 to 2025 A.D. Impressive achievement in last 15 years can be visualized with the fact that, in 1970, ONGC started offshore drilling, firstly near Aliabet in Gulf of Cambay.

In early 1974, the Sagar Samrat drilling platform at Bombay High, acquired from Japan, started functioning, reaching production of more than 230.95 million tonnes in 1980, thus fully realizing the contribution of offshore oil reserve in national economy. In the field of natural gas, world proven reserve reached 90,325 billion cubic metres as per January 1, 1984.

World commercial production of gas in 1983 reached 1584 bcm. In India, the total reserve of natural gas according to 1980 estimates was 351.91 bcm, the offshore share being 270,96 bcm. 1982 total production was around 24.12 bcm with respective states contributions viz. Gujarat (48%), Assam (31%) and Bombay (21%).

The 1980 and 1981 productions were respectively 19.25 and 19.97 bem. Location wise, natural gas (chiefly marsh gas with some other gaseous hydrocarbons) usually accompanies the petroleum accumulations, thus supplying the propulsive force which carries the oil from the oil-sands into the wells and if the pressure is sufficient, brings the oil up to the surface.

Since gas is both essential for the production of the oil and is also valuable as source of fuel on the oilfields, care is being taken now to prevent the huge waste of gas which was formerly so common. The country is fortunate to have discovered natural gas offshore the Andaman and Nicobar Islands, west coast of the Gulf of Cambay and Kutch and in basins of Brahmaputra, Godawari and Cauvery Basin. Principal present offshore supply being the Bombay High, producing 270.96 bcm. But Gujarat coast was on top in production of natural gas utilized between 1976-1980 i.e. 6.95 bcm, the others in series being Assam and Bombay High with value 4.55 bcm and 3.12 bcm respectively, making a total Indian production of 14.62 bem.

However, the national objective, to be satisfied, is in need of compact and accurate programs with more developments in seismic surveys and interpretation capability, development of suitable shallow water indigenous mobile drilling platforms, much advancement in capability for design, fabrication and installation of fixed production platform, the same in the case of submarine pipelines, marine equipments, instruments, and sophisticated research vessels for further study, research, and resource development.

Seafloor deposits, within the EEZ jurisdiction of India, fit for present commercial production, comprise only of sands, gravels, corals, lime, shells and relatively small quantities of tin, titanium and iron. However potential seafloor mineral resources are, immense and includes both oozes and clays, phosphorites and manganese nodules, the rich muds and brines (found in oceanic rift areas) and the newly discovered polymetallic sulphides and cobalt crusts (in relatively deeper oceans). Sulphur is also being exploited but on a minor scale.

Continental shelf rocks contain a variety of hard minerals, from coal and iron to tin, but only small quantities are being extracted commercially. For all this, ocean mining has to cope with availability, accessibility, advisability, and legality, among other factors. Although Asia remains the continent with overall greatest promise of deposits of these valuable minerals, the superficial mineral deposits of Indian Ocean includes a wide variety of terrigenous, biogenous and authigenic mineral deposits.

The beach and offshore place deposits of the Indian Ocean may be some of the largest in the world. The biogenous deposits comprise of the corals on shallow banks and on the continental shelves and the oozes in the deep sea. The authigenic deposits consists of the phosphorites, found both along continental margins (S. Africa and western India), and around seamounts in eastern and western Indian Ocean.

The continental margins of S. Africa, east Africa, southern Arabia, western India and the Andamans are marked by strong upwelling and provide nondepositional environments which are conducive to the formation of phosphorites. The polymetallic nodules found mostly in Central Indian Ocean Basin, beyond the national EEZ limit, cover an area of 10-15 × 106 km^2 and the resources are estimated to be about 1.5×10^{11} tonnes.

In recent past, the exploration and exploitation of mineral resources of India's EEZ have attracted considerable attention. Until the 1960's, the Indian Ocean was the least studied of the world oceans. It was mainly the International Indian Ocean Expedition of 1960-65, organized by UNESCO and ICSU, which considerably enhanced our knowledge of the ocean and mineral resources of this hitherto unexplored region.

Recent exploration of mineral resources of lesser value, in shallow offshore areas, has actively started, seeing its growing importance. The prospects of offshore mining in India's EEZ were studied in detail in 1977. By now, survey of her total EEZ is supposed to be complete with maximum required oceanographic informations, later helpful in exploitation activity.

Tertigenous deposits: Occurrences of heavy mineral concentrates usually known as "black sands" are reported from many localities of Indian coast, which contain ilmenite, rutile, zircon, magnetite, monazite, garnet, kyanite, and tin in significant proportions. The offshore extensions of some of these placer deposits have been explored. But only offshore tin placers are being exploited at present.

Most of the terrigenous deposits occur along the coast of Kerala between Quilon and Cape Comorin. The deposit is reported to contain 17 m tonnes of ilmenite, 1 m tonnes of rutile, 1.2 m tonnes of zircon, and 0.2 m monazite tonnes of. Ratnagiri coast (Maharashtra), Tinnaveli, Ramnad, Tanjore (Tamil Nadu), Visakhapatnam, Yarada, Waltair, Bhimunipatnam (Andhra Pradesh) and the coastal areas off Orissa.

Although the 25 km stretch along the coast from Neendakar to kayamkulam of Kerala state is supposed to have the richest deposit of offshore heavy minerals. Ilmenite concentration ranges from 17 to 79% in the area of Ratnagiri coast, forming an estimated reserve of 4×10^6 tonnes.

Further exploration of offshore placers will be guided by the proximity to onshore placers and examination of the relict sands on the middle and outer shelves. The present data indicate that along much of the western shelf, from the Gulf of Kutch i.e. Vengurla to the vicinity of Mangalore, the relict sands consist of oolites and biogenous materials. The extensive zone of potentially rich sands continues east to the Gulf of Mannar. However, it needs careful examination before starting the extensive exploitation.

Worldwide offshore mining of sand and gravels currently exceed the offshore productions of all nonfuel minerals in both value and volume. But in the Indian case it has to compete with onshore deposits, mostly near coastal urban centres viz. Cochin, Madras, Visakhapatnam and Calcutta. But at the same time, the growing deepening of the sea near the land, is likely to increase coastal erosion.

Biogenous deposits : Besides shell deposits of Crassostrea madrasensis, C. cucullata, Villorits cyprenoides, Modiolus sps. and Metritrix sps., the other major biogenous deposits are the corals and the deep see pelagic oozes. The best known deposits are in the Gulf of Kutch and in the Vambanad lake (back waters) of Kerala. Production from both these days being more than 1 million tonnes.

Lagoons of the Laccadive Islands, the Andaman Islands shelf area and in the Gulf of Mannar and the Palk Strait appear to be promising. In general, an examination of the bottom sediment map within EEZ of the seas bordering India shows relict oolite and biogenous sands on the middle and outer shelves in the western coast continuously from the Gulf of Kutch to the vicinity of Mangalore and in patches along the eastern coast.

Biogenous residues and the calcareous sands, amounting to 1135 m tonnes around Laccadive Islands can be economically attractive when exploited. Regarding ornamental and semiprecious stones, India is becoming increasingly important as a supplier of marine ornamental shells and the exports have risen from 20 tonnes in 1969 to 466 tonnes in 1979. Further, India is yet not reached exploitation up to the optimum level, thus the way is still open for more exploitation and exports.

Authigenic (Chemogenous) deposits: The chemogenous deposits, formed due to rapid chemical reactions on the ocean floor in the vicinity of t' • Indian coasts, are found in the form of barium nodules off the west Andaman and the Laccadive Island coasts, phosphorite nodules off the north Andaman, and polymetallic nodules in the Central Indian Basin, of course beyond EEZ area bur under exploitable jurisdiction because of India being a pioneer investor.

In the area of production and reserves of phosphate rock, India is holding 13th position in the world, with her 1980 production measuring 0.6 million tonnes out of an estimated reserve of 108 million tonnes. This phosphate deposit is marked with vital presence of valuable uranium, 100 to 200 parts per million. Algal and oolitic limestones, are also present off Saurashtra through Goa to Kerala coast.

Further India's agro-dominated economy needs phosphorous as major fertilizer, thus making offshore exploration for phosphate deposits unavoidable in near future. Gypsum crustals are found in the inner shelf off the Maharashtra coast between Vengurla and Bombay in the form of silty clay or clayey silt.

In general, a partial survey of EEZ of India has established that the offshore extension of heavy mineral placers on the western shelf are much larger than those available on land territory of India and coastal areas of other neighbouring countries.

Further large areas of the seabed in the India's EEZ, both deep and shallow, are covered by biogenic sediments, but their economic viability is still in doubt. But for islands like the Andaman and the Nicobar and the Laccadive, whose economies are becoming increasingly dependent on tourism, restraint would have to be exercised on the ruthless extraction of available mineral resource deposits. Systematic exploration and exploitation of Indian coasts for the phosphorite deposit is going to be indeed important in near future for purpose of self-sufficiency in fertilizers. Exploration of India's EEZ is nearing completion now.

Much emphasis should be placed on analysis and interpretation for greater understanding of the nature of deposits and technology needed for its proper exploitation. Besides exact location of valuable mining sites and detailed worldwide study of market position of these derived minerals, both in terms of physical availability and economic valuation, is must before undertaking any major capital intensive mining business in hand.

In fisheries, India occupies the 8th position in the world in terms of total annual catch, however the amount is the least in the world when calculated per fisherman engaged in this activity. Regarding the need, based on futuristic studies, the projected fish requirements of India's population by the year 2000 A.D. is estimated to be 11.4 million tonnes. Her EEZ is still half-heartedly utilized as proved by the fact that 50.6% of total annual catch of over 4.47 million tonnes from the Indian Ocean is harvested from the narrow coastal zone only, i.e., area lying between the shore and 28 meters depth range. The zone lying between 50 meters and 200 meters depth line contains 38.3% of total marine potential, while only 11. 1% of the total potential is there in the depth beyond 200 meters, i.e., 11.6% and 79.4% of area of total EEZ respectively.

In other words 9% of the inshore zone of EEZ produces more than 50% of the fisheries. Further the fisheries potential of the Indian Ocean is estimated at 14.4 million tonnes, of which the demersal fishes account for 7.4 million tonnes, the shoaling pelagic fishes for 6.0 million tonnes, the large pelagic fishes for 0.7 million tonnes, and the crustaceans and squids for 0.25 million tonnes. Prawns are found mostly within 50 meter depth area of EEZ, the only known exceptions being off the south western coast in depths of 275 to 700 meters. Similarly a large trawling ground of deep water lobsters, discovered beyond the 50meters depth, was located between Quilon and Cochin.

Tuna, having a wide migratory range, has been traditionally harvested from India's EEZ around the Minicoy and Laccadive Islands, although at lower levels.. Approximately 60% to 70% of the high priced tuna species of the Indian Ocean were taken by three non-Indian Ocean States namely Japan, Taiwan and South Korea. In 1976 India caught approximately 7-10% of the tuna and tuna-like fishes in the Indian Ocean.

Yellowfin and skipjack are the two major species of tuna found within and close to the Indian EEZ. Being the only underutilized tuna variety, skipjack, is in commercially exploitable potential in the Indian Ocean. India's catch of skipjack from Minicoy and Andamans through live bait method has gone up since 1973.

Besides tuna fishing requires highly specialized and technologically sophisticated operations in which India has little experience. Joint ventures and increased transfer of technology in the field of tuna fisheries is the need of the time.

Production through aquaculture has been increasing rapidly in recent past which can be evidenced with the fact that in 1980 its production was estimated to be 12.1% of the world's nominal catch, while in 1975 it was only 8.6%. India also leads in fin fish production in the world with an annual production reaching approximately 830,201 tonnes in 1981. But the fact remains that still 65% of the total Indian marine fisheries comes through traditional methods of fisheries only.

A comparative study shows that marine fisheries production in India has increased from 0.55 million metric tonnes in 1951 to over 1.6 million metric tonnes in 1980. On the economic front it contributes only about 0.5% of the GNP and 2% of the total export earning per year which, while considering its vast EEZ area, seems to be a very meager amount.

Of the total annual catch of around 2.5 million metric tonnes in 1980, 1.6 million metric tonnes were marine products; the aquacultures have been 850,000 metric tonnes, i.e., around 50% of the total marine products. However, by the year 2000 A.D. India could be producing 7 million metric tonnes through aquaculture. In the case of field prawn production in 1982, about 30,000 hectares of brackish water have been demarcated. Under government sponsored plan it is proposed that at least around 5,000 hectares should be added annually for prawn culture in marine water. Thus in the field of mariculture an estimated 1.7 million hectares of estuarine or brackish water areas and 8.9 million hectares of inshore waters are to be utilized along the Indian coast within its EEZ limit.

Finally for the successful resource management in Indian EEZ, the Indian government has recognized the main thrust areas for the optimization of investment at present, mainly high priority areas, manpower development has been sought at higher skilled levels which would successfully result in improving socio-economic condition of the vast population living mainly on the sea cost of the country.

Steps to start political and legal shells to understand the application of various process of UNCLOS III, mandatory transfer of technology and other associated complex issues like conservation, boundarly demarcation, pollution and protection of nation's interests in the EEZ are being sincerely undertaken. Comparative cost analysis in exploitation of these living and non-living resources is being seriously studied and their strategic importance being realized.

However, the important considerations which need immediate attention regarding resource development in Indian EEZ are:

(1) clarification of the level in concrete statistical terms up to which present EEZ utilization is going on;

(2) demarcation of optimum limit of EEZ utilization both in terms of exploration and exploitation taking an exact account of maximum sustainable yield;

(3) projected need of marine substances both living and nonliving of the EEZ by 2000 A.D. by more realistic evaluation of both protein requirements and energy substitution in different sectors;

(4) technological requirement for the fullest exploitation of EEZ resources and extent of foreign collaboration or technology transfer to be required by 2000 A.D.;

(5) planning of offshore construction and operational structure in total and the investment to be incurred by 2000 A.D. for the optimum resource generation from India's EEZ.

5 SEA LEVEL RISE

People have had the opportunity of extensively develop the shorelines of the world, because for the last several thousand years, sea level has risen so slowly that for most practical purposes it has been considered constan. Whether one is talking about a vacation sport in Rio de Janerio; swamps in Bangladesh, farmland in the Nile Delta, marshes along the Chesapeake Bay, or the Merchants of Venice, life along the coast is in a sensitive balance with the level of the sea. This balance would be upset by the rise in sea level that could result in global warming, which could raise sea level by one metre or more in the next century, due to expanding ocean water, melting mountain glaciers, and perhaps eventually causing polar ice sheets to melt, or slide down into the oceans. Sea level rise would inundate low-lying areas, drown coastal marshes and swamps of rivers, bays and aquifers throughout the world.

The global average rise of sea level depends primarily on (a) the shape and size of ocean basins, (b) the amount of water in the oceans, and (c) the average density of sea water. Subsidence, emergence and other locals can cause trends in 'relative sea level' at particular locations to differ from trends at "global sea level". Hays and Pitman analysed fossil records and concluded that over the last 100 million years, changes in mid-ocean ridge systems have caused sea level to rise and fall by over 300 metres. However, Clark (et.al) have accounted for sea level change of less than one millimeter per century. The impact of climate on sea level has been most Pronounced.

Geologists generally recognize that during ice ages the glaciation of substantial portions of the Northern Hemisphere has removed enough water from the oceans to lower the sea level one hundred metres below the present level, i.e., during the last (18,000 years ago) and previous ice ages.

Although the ice once covering much of the Northern Hemisphere have retreated, the world's remaining ice covers contain enough water to raise the sea level by over 75 metres. Estimates show that existing alpine glaciers contain enough water to raise the sea level by 30 or 60 centimeters. The Greenland and the West Antarctic ice sheets, each contain enough water to raise the sea level by about 7 metres, while East Antarctica has enough ice to raise the sea level by over 60 metres.

However there is no evidence that either the Greenland or the East Antarctic ice sheets have completely disintegrated in the last two million years. However, it is generally recognized that sea level was about seven metres higher than today during the last interglacial period, which was one to two degree centigrades warmer.

Tidal gauges have been available to measure the change in sea level at particular locations over the last century. It was found that the rate of sea level rise for the last fifty years would be between 2.0 and 2.5 mm/yr, while in the previous fifty years there was little change. Several researchers have attempted to explain the source of current trends in sea level rise. Barnett and Gornitz, Labedeff and Hansen, estimate that thermal expansion of the upper layers of the oceans resulting from the observed global warming of 0.4°C isotherm had migrated 100 metres downward, and concluded that the resulting expansion of ocean water could be responsible for some or all of the observed rise in relative sea level. Meier estimates, that the retreat of alpine glaciers and small ice caps could be responsible for a current contribution to a sea level of between 0.2 and 0.72 mm per year.

Concern about a substantial rise in sea level as a result of the projected global warming stemmed originally from Mercer, who suggested that the Ross and Filchner-Ronne ice shelves might disintegrate, causing a deglaciation of the West Antarctic ice sheet and a resulting six to seven metres, rise in the sea level in possibly 40 years.

Subsequent investigations have concluded that such a rapid rise is unlikeiy. Hughes estimated that such a disintegration would take at least 200 years, and Bentley estimated that it would take even more than 500 years. Hoffman, Wells and Titus estimated in 1985, that a warming of between 1° and 2.6°C could result in a thermal expansion responsible for contribution to sea level of between 12 and 26cm by 2050 respectively. In 1984, Meier estimated that a 2.8 mm rise had resulted from warming of 0.5^°C and concluded that a 1.5°C to 1.5°C warming would result in a rise of 8-25 cm in the next century.

Available estimates of the Greenland contribution assume that all meltwater flows into the oceans and that the ice dynamics of the glaciers do not change. The NAS Polar Board in 1985 suggested that some of the water would refreeze decreasing the contribution to sea level rise, although a change in the dynamics might imply additional deglaciation and increase the rate of sea level rise. The panel assumed that such changes were unlikely to occur in the next century. The potential impact of global warming on Antarctica in the next century, is the least certain of all the factors by which a global warming might contribute to sea level rise. Meltwater from East Antarctica might make a significant contribution by the year 2100.

Although most attention has focused on projections of global sea level, impacts on particular areas would depend on local relative sea level. Tidal gauge measurements suggest that relative sea level has risen 10 to 20 cm per century, more rapidly than the worldwide average along much of the U-S-coast. Local subsidence and emergence are caused by a variety of factors. Rebound from the retreat of glaciers after the last ice age has resulted in the emergence of Alaska and parts of Scandinavia.

Groundwater pumping has caused rapid subsidence around Houston, Texas, Taipai, Taiwan, and Bangkok, Thailand, among other areas. River deltas and other newly created land subside as the unconsolidated materials compact. Although subsidence and emergence trends may change in the future, particularly where anthropogenic causes are curtailed, no one has linked these causes to the future climate change in the next century. However, the removal of ice from Greenland and Antarctica would deform the ocean floors. Other influences on local sea level that might occur as a result of a global warming include currents, winds, and freshwater flow into estuaries. None of these impacts, however, has been estimated.

A rise in sea level of 1 or 2 metre would permanently inundate wetlands and lowlands, accelerate coastal erosion, exacerbate coastal flooding, threaten coastal structures, and increase the salinity of estuaries and aquifers.

The most direct impact of a rise in sea level is the inundation of areas that had been just above the water level before the sea rose. Coastal wetlands are generally found at elevations below the highest tide of the year and above mean sea level.

Estimating past sea level rise has been the analysis of marsh peats commonly. Thus the impacts of sea level rise on wetlands are fairy well understood. For the rates of sea level rise of the last several thousand years, marshes have generally kept pace with sea level through sedimentation and peat formation.

As sea level rose, new wetlands formed inland while the seaward boundary was maintained. Because the wetland area has expanded one would expect a concave marsh profile, that there is more marsh area than the area found immediately above the marsh. Thus, if sea level rose more rapidly than the marsh's ability to keep pace, there would be a net loss of wetlands. Moreover, a complete loss might occur if protection of developed area prevented the inland formation of new wetlands.

Throughout the world, people have dammed, leveed and channelized major rivers, curtailing the amount of sediment that reaches river deltas. Even at present rate of sea level rise, substantial amount of land is converting to open water in Egypt and Mexico.

Other deltas, such as the Ganges in Bangladesh and India are currently expanding seaward. These areas would require increased sediment to keep pace with an accelerated rise in sea level. Additional projects to divert the natural flow of river water would increase the vulnerability of these areas to a rise in sea level. Several options have been identified for reducing wetland loss due to sea level rise.

Abandonment of developed area inland or existing wetlands could permit new wetlands to form inland. In some cases it might be possible to enhance the ability of wetlands to accrete vertically by spraying sediment on them or restoring the natural processes that would provide sediment to the wetlands.

Although coastal wetlands are found at the lowest elevations, inundation of lowland could also be important in some areas, particularly if sea level rose at least by one metre.

Unfortunately the conventional ten feet contours in used the mapping of most coastal areas has prevented a general assessment of land loss. Analysing date from aerial photographs to assess elevation in the area around Charkeston concluded that 160 and 230 cm rises could result in 30 per cent and 46 per cent losses of the areas dry land respectively. Such rises would result in 9 per cent and 12 per cent losses of the land in the area of Galveaton and Texas city.

As with wetlands' loss the responses to inundation fall broadly into the categories of retreat and holding back the sea. Levees are- used extensively in the Netherlands and New Orleans to prevent the flooding of areas below sea level and could be similarly constructed around other major cities.

In sparsely developed areas, however, the cost of a levee might be greater than the value of the property being protected. Moreover even where levees prove to be cost effective, the environmental implications of replacing natural shorelines with man-made structures would need to be considered.

Sea level rise can also result in the loss of land above sea level through erosion. In 1962 it was shown that the erosion resulting from a rise in sea level would depend upon the average slope of the entire beach profile extending from the dunes out to the point where the water is too deep for waves to have a significant impact on the bottom (generally a depth of about 10 metres.)

By comparison, inundation depends only on the slope immediately above the original sea level. Because beach profiles are generally flatter than the portion of the beach just above sea level, if implies that the erosion from a rise in sea level is several times greater than the amount of land directly inundated.

Processes other than sea level rise also contribute to the erosion process including storms, structures currents and along shore transport. For example, water levels along the great lakes can fluctuate over one metre in a decade.

Geologists basically believe that coastal barriers can maintain themselves in the face of the slowly rising sea level through the landward transport of sand which washes over the island during storms, building the island upward and landward.

The potential erosion from a rise in sea level could be particularly important to recreation beach resorts, which include some of the world's most economically valuable and intensively used land. Relatively few of the most densely developed resorts have beaches wider than about 30 m at high tide.

Thus the rise in relative sea level of 30 centimeters projected in the next 40 to 50 years could erode most recreational beaches in developed areas, unless additional erosion response measures are taken.

Therefore, in many undeveloped countries, relatively inexpensive houses are found very close to the shore. Because the value of these houses is less than the cost of protecting them, they must be moved as the shore erodes.

An accelerated rise in sea level would speed this process of shoreline retreat.

A rise in sea level could increase flooding and storm damages in coastal areas for three reasons: erosion caused by sea level rise would increase the vulnerability of communities, higher water levels would provide storm surges with a higher base to build upon, and higher water levels would decrease natural and artificial drainage.

In 1984, Leatherman conducted an analysis of Galveston Island, Texas. He estimated that the area within the 100 year flood plain would increase from 50 per cent to 94 per cent for an 88 cm rise in seal level and that for a rise greater than one metre. The Galveston seawall would be overtopped during a 100-year storm.

A wide variety of shore protection measures would be available for communities to protect themselves from increased storm surge and wave damage due to sea level rise. Many of the measures used to address erosion and inundation including seawalls, breakwaters, levees and beach restoration also provide protection against storms.

In the case of Galveston which is already protected on the ocean side by the seawall, it might be necessary to completely encircle the developed areas with a level to prevent flooding from the bay side, upgrading the existing seawall might also be necessary.

Although most researchers and general public have focussed on the increased flooding and shoreline retreat associated with a rise in sea level, the inland penetration of salt water could be important in some areas. Besides as a rise in sea level increases the salinity of an estuary represents the outcome that (1) the tendency for the ocean salt water to completely mix with the estuarine water, and (2) the tendency of fresh water flowing into the estuary to dilute the saline water and push it back toward the ocean. During the rainy season, therefore, low salinity level prevails. A rise in sea level has an impact similar to decreasing the freshwater inflow. By widening and deepening the estuary, sea level rise increases the ability of saltwater to penetrate upstream. Hull, Thatcher and Tortoriello in 1986 examined the potential impacts of an accelerated rise in sea level due to the greenhouse warming. They estimated that 73 ern and 250 cm rises would result in ocean front migrating as much as additional 15 and 40 kilometres, respectively.

Thus the causes and effects of sea levels rise are a matter of grave concern and it, is now understood that if proper steps to restrict the rise are not taken, then dangerous consequences would result in a magnified way and affect the human life in a very horrific manner, because it is to be clearly understood that human life would also be affected by the rising sea level, especially of the people in the coastal areas. Thus it can be concluded that the problems of future sea level rise are important not only to geomorphologists, ecologists, geographers and social scientists, but also to governments and their agencies concerned with planning and management of the coastal regions that will be modified by the changes.

Of late, it has become increasingly clear that man has acquired the power and capability to alter the ecology of this planet adversely. Researchers in the past few decades have clearly demonstrated that urgent attention is needed to reverse this trend. The influence of the oceans which cover nearly 71 per cent of the earth's surface, is of much importance for regulating the climate of this planet.

The last decade has witnessed sudden increase in man's awareness of the oceanic environment, and the progressive deterioration of our seas due to human activity. Any change in the equilibrium of the sea, however minor it may be, is bound to affect the human population, as 20 per cent of the total global population is living in the coastal belt of the landmass.

Therefore, it is a matter of utmost importance to develop an understanding of the ocean and coastal environment. Moreover, the changes in the sea level is a subject which deserves our immediate attention. The global warming, an important factor for sea level rise, is caused by the release of carbon dioxide, chlorofluorocarbon, nitrous oxide and methane in the atmosphere due to human activity.

The consequences of sea level rise are far reaching and these result in increased coastal erosion, submergence of coastal wetland, increased damage due to storm surges, shifting of the river mouths and deltas, salt water intrusion into the rivers, etc. In addition, ecological damages include loss of coastal vegetation, loss of commercial fisheries, destruction of important nursery grounds like the mangrovers and extinction of valuable coastal fauna and flora.

The most severe impacts are likely to stem directly from inundation. Asia would be most affected because of the extreme vulnerability of several large and heavily populated deltaic regions. For example, with a 1.5m sea-level rise, about 15 per cent of all and (and about one-fifth of all farmland) in Bangladesh would be inundated and a further 6 per cent would become more prone to frequent flooding.

Altogether 12 percent of agricultural production could be lost. In Egypt, it is estimated that 17 per cent of national agricultural production and 20 per cent of all farmland, especially the most productive farmland, would be lost as a result of 1.5m sea-level rise. Island nations, particularly low-lying coral atolls, have the most to lose. The Maldive Islands in the Indian Ocean would have one-half of their land area inundated with a 2 inch rise in sea-level.

In addition to direct farmland loss from inundation, it is likely that agriculture would experience increased costs from salt water intrusion into surface water ground water in coastal regions. Deeper tidal penetration would increase the risk of flooding and rates of abstraction of ground water might need to be reduced to prevent recharge of aquifers with sea water.

Further, indirect impacts would be likely as a result of the need to relocate both farming populations and production in other regions. In Bangladesh, for example, about one-fifth of the nation's population would be displaced as a result of the farmland loss estimated for a 1.5 m sea-level rise.

Historically, the coastal regions around the world, particularly areas near the mouth and along the great rivers have been centres of population concentrations. Trade, commerce and industry have flourished well because of easy access to the resources of the hinterland and reach of the markets across the oceans.

Thus, it is not surprising that a significant portion of the world's population lives within the coastal zone with many of the cultural artifacts, built at elevations less than 3 m (10 ft.) above mean high-tide level along the shoreline. Even without a change in the height of mean sea level or the height of the land, the elevations of settlements and other artifacts are not adequate to provide safety in the event of extreme stormsurges, etc.

The hazard of flooding, erosion, intrusion of saline water in the underground aquifers has grown increasingly apparent and serious along many of the world's coastline as local mean sea level has risen during the twentieth century at a rate ranging from 1 to 5 mm/yr. The scientific and engineering community has focused their attention of the problems of accelerated beach erosion and the rising level of green house gases in the atmosphere."

As a first approximation, the effects of one metre sea level rise on a coastline in South Asia may be determined by surveying the contour one metre above present high spring tideline, but allowance must then be made for the effects of erosion and deposition as submergence proceeds.

Submergence will result in recession of the coastline, except where there is vertical cliff, or where deposition of sediment continues at a sufficient rate may continue in the civility of the mouths of major sediment-yielding rivers, such as the Ganges, which have built large deltas.

Submergence are likely to accelerate the recession of cliffed coasts, except where the rock outer crops are resistant, and high and low tide lines simply move up the cliff face. In the tropics cliffs are mainly on headlands, such as Cape Comorin in India and Condra Head in Sri Lanka, and the rock outcrops are usually deeply weathered and easily eroded.

Many are vegetated bluffs rather than actively receding cliffs, and a sea level rise is likely to understand them, producing slumming and eventually steep, active cliffs. Cliffs that are already receding, as on the coast near Chittagong and Cox's Bazar in Bangladesh, will show an acceleration of cliff retreat as sea level rises; and larger waves move in through deepening water.

Where shore platforms have developed, they will disappear beneath the rising sea. Some will be built up by nearshore deposition, or the growth of organisms such as algae and corals. Submergence of low-lying rivers deltas and coastal plains will cause extensive retreat,, of the coastline, accompanied marine erosion, but a few sectors may be maintained by sedimentation, especially in the civinity of the mouths of large rivers.

On the north coast of Java there is already erosion on deltaic sectors where former river mouths have been diverted. Naturally or artificially, but this is a largely compensated by the provocation that continues around the present river mouths. With a rising sea level, the proportion of eroding coastline will greatly increase on these and other deltas.

In India, Ahmed estimated that a quarter of the coastline was eroding as the result of delta growth, but this proportion will be much reduced. As sea level rises, beaches will become narrower, and beach erosion will become more extensive than it is now. An exception may be found locally where beaches are nourished with a larger supply of sediment derived from rivers draining areas of increasing rainfall, or from accelerating clifferosion, but some of the sediment supplied to the coast by rivers will be retained within submerging river mouth areas.

As submergence proceeds, coastal inlets, embayments, and estuaries will enlarge and deepen and salinity penctration from the rising sea will increase. Tidal regimes will be modified, and intertidal shoals, mangroves and marsh islands (as at the mouth of the Ganges) will show' extensive erosion as these features disappear beneath the sea.

Depositional terraces built up by mangroves and salt marshes during the Holocene slits and will only be maintained if sedimentation continues at a sufficient rate, but generally their seaward margins will be cut back.

Coastal lagoons of the kind found in Southern India and around Sri Lanka sea will erode their enclosing barriers in some cases submergence and erosion of barriers will reopen lagoon and coastal inlets or embayments.

Fringing and nearshore coral reefs will be submerged in the name way as rocky shore platforms fronting cliffed coasts, but the rising sea level may stimulate the upward growth of corals and associated organisms, which could maintain the reef structure if they keep pace with the marine trans region. This will be critical for the future of low sandy islands (cays) which have formed on coral reefs flats as in the Maldives and the Andaman islands. Low-lying coastal areas will show migrations of vegetation and faunal zones as submergence proceeds.

Mangroves are extensive in such areas as the Sunderbans and the tidal shores of the Ganges delta, but mangrove fringed coastlines have become less extensive in Southeast Asia in recent decades because of the impacts of land reclamation, fishpond construction, mineral dredging, and waste disposal. In Thailand, for example, these impacts have already reduced the mangrove area by more than one-third.

Where mangroves persist, they typically stand in front of zones of salt marsh and freshwater vegetation (for scrub or swamp forest) where there has been coastal degradation, the advance of the mangroves (often arranged in species zones) has been followed by salt marsh and fresh water communities in succession as sedimentation builds up the substrata to appropriate levels, forming an intertidal depositional terrace.

A sea level rise will reverse this sequence, unless sedimentation continues at a sufficient rate to maintain or propagate the coastline. Submergence will kill the sewad mangroves, and initiate erosion of the previously prograded terrace. As the coastline retreats, the mangrove zone will migrate landward to displace the salt marsh, which in turn will invade the freshwater hinterland. Mangroves regenerate quickly in areas that have been cleared, then abandoned, and it is likely that they will spread back on to suitable low-lying hinterlands a's the sea rises, but where the hinterland is steep or rugged the mangrove zone will be extinguished by submergences.

In South and South-East Asia, the land immediately behind the mangroves is commonly used intensively for fish ponds or rice fields and these would be invalid by the retrogressing mangroves if local people permit. In drier areas, such as the Rann of Kutch where mangroves are backed by base, hypersaline plains, a sea level rise will result in the colonisation of presently unvegetated tracts by mangroves and the coastline retreats. Coastal lagoons typically show salinity gradients from fresh river inflow through brackish to sea water at their marine entrances; this is true for example in the chain of lagoons that opens to the sea at Songhkla in South Eastern Thailand.

A sea level rise of a metre over the next century is likely to increase marine inflow through deepened and sidened marine entrances. Where this occurs areas that were previously fresh water will become blackish with dieback of existing vegetation and replacement by mangroves, salt marsh and other salt tolerant hydrophytes. There will also be displacement of fresh water fauna by estuarine and marine species.

In these various ways ecosystems in coastal regions will change as sea level rises on the coasts of South and South East Asia. The natural vegetation and fauna have already been extensively modified and what actually happens will depend very much on human responses to ecological change.

The way in which modern coastal societies will respond to be a relatively rapid rise in sea level will vary with political and economic factors. Nations that have the organization, technology and resources to counter the effects of a sea level rise by means of artificial structures will do; so others will, in general have to choose between evacuation of submerging areas or adaptation to changing coastal environments.

On coastline close to urban and industrial centres it is likely that a sea level rise about a metre over the next century will stimulate expenditure on structures designed to prevent submerging and erosion. It is very likely that where land has recently been reclaimed from the sea for use by large coastal population (as in Singapore), strenuous efforts will be made to retain it, by building sea walls and introducing pumping systems using the techniques familiar from the history of the Netherlands coasts. Where Erosion Beaches are valued for their creation or tourism, as in seaside resort such as Pattaya in Thailand it is possible that they will be maintained by beach refreshment programmes of the kind already used in the US and Australia.

More generally beach erosion will continue to be met by the spread of artificial structures; already more than 260 kms. of the Kerala coastline in South India are fringed by granite boulders walls.

Away from highly developed centres there may have to be abandonment of coastal fringes as submergence proceeds and modification of developed land and water uses as the water table rises and soil and water salinity increase. A sea level rise of a metre over the next century will accentuate such problems on the intensively utilised and densely populated coastal plains of South and South-West Asia, causing coastline recession of up to several kilometres, displacing coastal villages and depriving many people of their land and resources. Tide range in right of Bangkok, is up to 2.3 metres and a sea level rise of a metre point gently sloping coastal plain would submerge the whole of the reclaimed coastal plain would submerge the whole of the reclaimed mangrove area and the additional zone up to five kilometres to landward existing fish ponds, shrimp ponds and salt pans are thus likely to be submerged and destroyed by marine erosion unless they are protected by even larger and more elaborate sea walls.

The predicted sea level rise of a metre during the coming century will greatly modify the geomorphology and generate a whole new range of problem for the communities that occupy and utilize these coastal areas.

The effects of a sea level rise can be reduced theoretically or inferred from studies of coastlines where submergence is already in progress, but there is a need for site specific predictive studies of the physical and ecological consequence of a sea level rise on the various types of coastline, especially those that are low lying densely populated and intensively utilized.

The human response to sea level may be counter measures, where resources and technology are capable of these but elsewhere there will have to be evaluation or some kind of adaptation to the changing coastal environment. On the coastlines of South and South-East Asia, the most likely outcome is an intensification of use and occupancy of near shore sea areas.

Global climate change when talked of brings about many such changes, which directly or indirectly affect human activities, the social, economic and political set-up. This sea level size directly effects the life of human beings, especially those, who live in coastal areas. Their economics and politics are very much dependent on the sea, and any changes in the sea will naturally bring about changes in their life-styles, economics and politics, equally, as all three are interrelated. Ecologically, sea level size, brings about consequences like the changing of the area's biosphere, which would naturally affect human activities, as the people living near the sea, would be very much dependent on the biotic resources, especially those obtained directly from the sea, like fish, and other such sea foods, pearls etc.

Submergence of coastal wetlands, inundation, coastal erosion, flooding and storm damage, especially in the coastal erosion, flooding and storm damage, especially in the coastal areas, increased salinity in estuaries and aquifers, and all these factors would affect the human life to such a degree that, special efforts are required to right these consequences, brought about by the increasing sea level.

As a consequence, the coastal zone is to be regarded as one of the most important and critical parts of the planet, particularly as far as global change is concerned. On the one hand, the physical context is transforming because of changing sea level and progressive coastal erosion; on the other hand human communities concentrating along the sea exploit its coastal resources.

There is a growing awareness today that human activities are beginning to affect out plant on a global scale. Sea level rise and other potential consequences of climate change create problems that can be solved only by international cooperation.

There is an interaction between social conditions and the impacts of sea level rise. On the one hand, particularly if coastal defence is impossible there could be a large social impact in the form of the abandoning coastal areas and the migration of population whether or not this is considered acceptable, depends on the values and norms of the society. On the other hand, the way in which a society responds to threats from a rising sea level depends very much on the social structure. Seventy per cent of the human population lives within 60 km of the sea, and the percentage is increasing. Human settlements, industry and agriculture in coastal lowland are vulnerable to rising sea level or changes in storminess or storm surges.

Some nations are particularly vulnerable. Eight to ten million people live within one metre of high tide in each of the unprotected river deltas of Bangladesh, Egypt and Vietnam. Half a million people live in archipelagos and coral atoll nations, that lie almost entirely within three metres, of sea levels, such as the Maldives, the Marshall Islands, Tuvalu, Kiribati and Tokelau.

Other archipelagoes and islands nations in the Pacific and Indian oceans and Caribbean could lose much of their beaches and arable lands, which would cause severe economic and social disruption. Even in nations that are not, on the whole, particularly vulnerable to sea level rise, some areas could be seriously threatened. Examples include Sydney,

Shanghai, coastal Louisiana and other areas economically dependent on fisheries or sensitive to change in estuaries habitats.

As a result of present population growth and development, coastal areas worldwide are under increasing stress. In addition, increased exploitation of nonrenewable resources in degrading the functions and values of coastal zones in many parts of the world. Consequently, populated coastal areas are becoming more and more vulnerable to sea level rise and could have serious adverse effects. A rapid rise of the mean sea level 1 to 2 metres in 100 years will have a considerable impact on all environmental conditions in low deltaic areas, coastal marshes and embayments in coastal lowlands. They are often densely populated, and highly productive in the agricultural sector. The impacts on morphology and water quality will lead to counter measures in the fields of flood protection and water management.

A common situation is that of a growing delta. The land areas are built up by sediments from the river and the sea, to an elevation slightly above high tide in the coastal strip and to the level of 1 to 5 years flood along the rivers. So the land level rises and the delta acerbates into the sea. The whole process is strongly related to the local relative change of mean sea level.

A rapid increase of the global rise of mean sea level will upset the present dynamic equilibrium. The process will adapt to the new rate of rise which generally means a slowing down of the rate of accretion even recession and drowning may be the result. A rise in mean sea level will also influence the propagation of the tides and the generation of storm surges.

Sea dikes will have to be high often even more than the rise of mean sea level. In some cases a shortening of the coastline by enclosure of embayments may be feasible. The rise of sea level and the related backwater effects increase the water levels in the estuary and, with some time lag, in the lower part of a river. In these areas also the floor protection has to be adapted.

From a geological point of view deltas belong to the most rapidly changing parts of the crust of the earth. A common situation in cases where man has not yet interfered or only to a limited extent is that of a growing delta. This is a delta which is extending into the sea entailing a rise of the riverbed and hence of the levels of the river floods. The land areas are built up by sediments from the river and from the sea. Under normal conditions the land in the costal strips build up to an elevation slightly above high tide level to approximately the level of the 1 to 5 years flood.

A rapid rise of the sea level within a certain period will upset the present dynamic equilibrium. There will be a gradual adoption of the morphology to the new sea level. Because of the time lag the water depths of the river channel and the sea floor will initially increase slowing down the rate of building up.

The tidal currents in the sandy coasts. The mangrove belt which exists along the disappear. This belt which exists along the coasts of most tropical deltas will deteriorate or even disappear. In deltas where the distribution of the upland discharge over the tributaries has not been stabilized by hydraulic works, a river branch may be cut off from the supply of sediments (morbund river). The part of the delta which was supplied with sediments by that branch will be exposed to erosion in the case of a rapid rise of the sea level (parts of the delta of the Mississippi River).

An extreme case is that of the drowning delta like the (former) delta of the Rhine and the Meuse in the Netherlands. Now supply of sediments, natural and man-induced subsidence and a rise of the sea level have led to an artificial situation where man is constantly fighting the destructive forces of the sea. Man has learned how to cope in this area with a continuing relative rise of the sea level which, if man induced subsidence is included, amounts to some 3 metres during the past 1000 years.

Coastal embankments are exposed to high water levels and, depending on the situation, to impact of waves. High sea levels are caused by astronomical spring tides and along a number of coasts, by wind-induced set-up of the sea level (storm surges, cyclones, typhoons and hurricanes).

A rise of the sea level will affect these conditions in different ways. The propagation of the astronomical tides in shallow coastal waters will be modified, increasing, in general, the tidal range. The set-up will decrease but, in general, this effect will be small. The wave uprush may substantially increase, especially in those cases where under the original conditions, the depths in front of sea dikes are small or where tidal foreland exists.

The result is that the sea dikes will have to be raised to a greater height than the rise of the sea level to maintain the same degree of safety. In many deltas embankments have been raised periodically to cope with the relative sea level rise.

The motive has often been not so much for this rise but the desire to ensure a greater degree of safety, especially after disasters. This was also the motive for the implementation of schemes aiming at shortening the lines of defence by enclosures near the mouth of estuaries and other tidal embayments.

In all estuaries where a fresh water river debauches into a basin with saline water the heavier saline water will penetrate into the estuary and more in an upstream direction in spite of the flow of freshwater in the opposite direction.

Depending on various factors, in the first place the tidal range in the basin, the estuary will show a distinct interface between the fresh water near the surface and the saline water near the bottom or a situation will occur when there is a partial or complete mixing. The saline intrusion renders the water unsuitable for various purposes and is a maximum when the riverflow is minimum during dry periods when the need for fresh water is high. An extreme case is the Gambia river in West Africa where at minimum flow the saline effect is felt up to 235 km. upstream from the sea.

A rise of the-sea level will increase the intrusion of saline water because of the increase in water depths, at least in a first phase, and the increase in tidal flood volumes. The shift of the saline reach in the upstream direction will affect the intake of water from the river for irrigation and water supply and the ecological conditions of the channel and the adjacent land.

It should be noted that this process is already taking place in a number of estuaries where because of dredging in connection with post development the saline effect has moved further upstream. A classical example is the Rotterdam waterway giving access to the port of Rotterdam where the saline reach moved over 35 km. in a period of about 60 years.

Low land rice is grown in many deltas in the humid tropical zone where storm surges occur and where there is no protection against river floods. In these areas, often there are only low coastal embankments to protect the coastal strip from flooding with saline water during high astronomical spring tides. Local rice varieties are grown which are adapted to the depth of flooding caused by river.

Near the coastal strip the depths of flooding are small and relatively high yields are obtained. Further upstream the depth of flooding maybe as much as 3-4m and floating rice' is grown with low yields. These conceptions occur in many parts of the large deltas of South East Asia and especially in the delta of the Mekong river.

A rise of the sea level of, say 1 m or more, would upset these conditions, the sea-dikes will have to be raised and in the tidal reach the depth of flooding will increase affecting the productivity. To maintain the present productivity levels protection against flooding will become necessary and embankments will have to be created.

Although, in itself, this is a simple operation, it will change the environment profoundly and will entail many hydrological and morphological side effects (rise of the flood levels, rise of the river bed and increased meandering). Embanking makes it necessary to lay out a system of drainage canals and outfalls. Because of the elimination of the beneficial effects of the floods (supplemental irrigation and flushing it may become necessary to provide irrigation and flushing) it may become necessary to provide irrigation facilities.

Under the present conditions with the continuing relative sea level rise and the desire to increase the yields and to grow other crops than rice there is a tendency to carry out development schemes in the deltas aiming at flood protection, drainage and irrigation. An uncertain economic feasibility often hampers actual implementation. A rise of sea level which far exceeds that occurring at the present will either ruin many agricultural areas or accelerate the developments which are already taking place.

Depending on the land elevation, relative to tidal levels, and on the drainage requirements (low and rice or dry food crops) either gravity drainage or pump lift drainage can be applied. In the coastal zone of many deltas all over the world tidal drainages is applied on a large scale when the land is not too low. The system often consists of simple wooden box sluices with glap gates.

If there is a significant rise of the sea level this system will have to be replaced by pump-lift. This will entail high installation and operation in deltas in the humid tropical zone with their high amounts of rainfall. Where pump-lift drainage is applied like in the Netherlands, Italy, Japan, USA and India the pumps will operate at lower efficiency if there is a substantial rise of the sea level and the structure may have to be modified or redesigned.

The sea level rise will increase the salt water intrusion threatening the intake points of irrigation water along the river. Unless the drastic solution of enclosure of the estuaries near the mouth is applied, the intake points will have to be shifted upstream beyond the salinity limit during periods of low river flow.

New canals will have to convey the irrigation water from the new locations of these intakes to the downstream areas and when the stops are small booster pumping stations in the canals will be necessary. A benefit of the sea level rise will be the increase in head for the flow of irrigation water from the river to the land.

If there is a rapid rise of sea level cities in low-lying coastal areas will have to be protected from flooding by encircling embankments (pocket dikes) and in any case, will have to be drained by pump-lift. This is an operation which is already under way in a number of large cities mainly as a consequence of man-induced land subsidence.

Typical examples are New Orleans and Osaka, Bangkok is a recent case. In Osaka a (local) maximum subsidence of 2.8 m occurred in a period of about 30 years. This probably is a greater rate than any conceivable rate of rise of the sea level. The walls along the city canals and the entrances from the sea have been raised by 1 to 3 m. check-sluices (some with vizor gates with a span of 60 m) have been built which can be closed when high typhoon levels occur on the Pacific ocean. The city area is then isolated from the surrounding waters and excess water from local typhoon rainfall has to be removed by pumping.

The Kema pumping station, probably the largest of the world, with a capacity of 330 m/s (pumps of 55m/s) and a head of 3-4 m has been constructed for this purpose. The response of the Japanese to this challenge has been immediate and vigorous. Because of financial and organizational problems this however would probably not be possible for cities like Shanghai and Calcutta, should a rapid relative rise of the sea level occur which would be similar in effect to the subsidence of Osaka.

The case of Osaka demonstrates that, from the technical point of view, area of the sea level of say 2 to 4m within a period of 100 years would not cause unsurmountable problems. The case has been justified because of the high economic value of an urban place like Osaka. In tidal harbour basins it will be necessary to raise the heights of quay walls and landing stages. Attention should also be paid to ensure that there is sufficient headway under bridges. In the case of dock harbours the access locks may have to be adapted to a rise of sea level.

In a number of coastal areas in France, India, Japan, the Netherlands and to a lesser extent Bangladesh and the Mekong-delta coastal or estuarine reservoirs have been framed by damming of tidal embayments or estuaries. Excess water from the rivers is discharged into the sea through sluices and the enclosed basins have been keen transformed into reservoirs with fresh water by the inflow of river water.

The discharge of to the sea is effectuated by tidal drainage. The low lying areas around the reservoirs can abstract water from these reservoirs during dry periods; generally they also drain to these reservoirs during rainy periods, either by gravity or by pump lift.

A rise of the sea level of around 0.5 m or more will completely upset the present operation of these reservoirs, since gravity drainage into the sea will no longer be possible. There are two basically different solutions which could be considered. The one could be, that the reservoir levels would be maintained at the original level. This implies that excess water from the reservoir would have to be removed by pumping and that the sluices would be closed. Since the volume of flood waters from the river may be quite high, this could be an expensive solution.

In the second solution operational level of the reservoir could be raised, so that gravity drainage to the sea remain possible. This would require modification or redesign of the existing sluices. The drainage of the surrounding low lying areas on the reservoir would be --hampered and pump-lift may be necessary.

Embankments would have to be raised not only along the reservoir- but, because the rivers debauching into the back-water effects, also along the rivers debauching into the reservoir. Indeed in a delta, with its system of interconnected watercourses, the side of the reservoir level would propagate in the entire system unless partitions are made. There would also be seepage flow from the reservoir with its high level to the surrounding lowlying areas, but after a long period of time this seepage water would have the same composition of the water in the reservoir.

A large population of the world lives in coastal areas and on islands. These areas have several assets, including some of the largest cities of the world, ports, industries, fertile lands, fisheries and tourism facilities. A sea level rise of one metre by 2100 would have severe impact on all coastal areas particularly in the developing countries. More frequent recurrence of storm surges would pose an intolerable burden on the coastal population, impact on shorelines erosion, inundation of wet lands, and mangroves, increases in the salinity of aquifers, and drowning of coral reefs would cause severe ecological damage.

The small islands are particularly vulnerable. Human beings in their responses would naturally try to adapt to these changes and the rising sea level especially in the coastal areas where the effect will be such higher. This adaptation will be brought about, because of the severe stress on the economies of those effected areas.

Agriculture will be affected as sea level area could inundate low lying coastal plains with fertile alluvial soils where a high proportion of rice production is located. Fisheries could be affected. Two-thirds of the world's fish catch and many marine species, depend on coastal wet lands for their survival.

Without human interference eco-systems would migrate landward as sea level rises and then could remain largely intact, although the total area of wetlands would decline and so would the fish catch. If human intervention is followed then there are chances, that a large portion of these ecosystems would be lost which again would reduce or finish the fisheries.

While it is very difficult to estimate the economic implications correctly, it to possible to make some broad statements about the various options. In densely populated and productive areas retreat may not be viable in general and in the case of small islands, could be impossible. Accommodation could be extremely costly but could be justifiable in particular situation. Protection would be economically viable, in particular cases but could be unaffordable for poorer countries. One estimate quoted in the Inter-governmental Panel on Climate Change report gives the total additional costs to be more than 500 billion dollars with a major part to be incurred in developing countries.

Thus a gentle coherence is needed and to understand the human response, they can be put into three categories.

(1) Retreat, i.e., abandon vulnerable areas and resettle-inhabitants.

(2) Accommodation—adjust to sea level rise

(3) Protection for vulnerable areas particularly those with high population, density, economic value or precious natural resources.

Retreat: Preventing development can reduce future expenditure foe adaptation. India, Sri Lanka, USA, Australia and a number of island nations have already such regulations. Where the area that could be immundated is high, the timing of such regulations would be difficult.

Spreading of awareness among all the occupants of vulnerable areas could encourage private initiatives and solutions. This is no solution, however for small island states and poorer countries. Also under this option, resettlement could create major problems. Resettled people are not always well received. They often face language problems, racial and religious discrimination and difficulties in obtaining employment. Even when they feel welcome, the disruption of families, friendships and traditions can be stressful.

Although the impacts of accommodation and protection would be less, they may still be important. The loss of traditional environments which normally sustain economics and cultures and provide recreational needs could disrupt family life and create social instability regardless of the response eventually chosen. Community participation in the decision making process is the best way to ensure that these implications are recognized.

Accommodation: This response requires advance planning. Many coastal structures could be elevated for protection from inundation. Building-codes could be devised for protection from surging waters and high winds. Drainage could be modified, storm warning and preparedness plans could be strengthened. Where salt water damages agricultural lands, salt-tolerant crops may be introduced. Some agricultural lands may be converted to agricultural uses.

Human activities that harm coastal resources maybe banned, e.g. filling wetlands, damming rivers, mining coral and beach sands and cutting mangroves. Undeveloped land can be set apart to accommodate re-establishment of wetlands and mangroves. Exploitation of ground water could be controlled to prevent salinity intrusion.

Protection: This response involves defensive measures against in addition, tidal flooding, shore erosion, salinity interaction and loss of natural resources. Those measures could be structures such as dikes, levees, flood walls, sea walls, and dune building, and pollution control.

Economic Implications: While it is very difficult to estimate the economic implications correctly, it to possible to make some broad statements about the various options. In densely populated and productive areas retreat may not be viable in general and in the case of small islands, could be impossible. Accommodation could be extremely costly but could be justifiable in particular situation. Protection would be economically viable, in particular cases but could be unaffordable for poorer countries. One estimate quoted in the Inter-governmental Panel on Climate Change report gives the total additional costs to be more than 500 billion dollars with a major part to be incurred in developing countries.

Social and Cultural Implications: The lifestyles of people living on the coasts are distinct. The adaptive measures may, therefore impact heavily on the social and cultural mores of millions of people. Relocation of people whether planned or unplanned, could be very high and cause great conflicts, even within the same country. Where such relocation crosses national boundaries, international conflicts could occur.

There is considerable optimism among many adoption to climate change is feasible, since it is as old as man himself. At every location, human beings and other living things adapted successfully to varying temperatures, precipitation, droughts, floods and storms. Because of economic and technological progress, the in-built adaptability of humankind has improved.

The world has many climatic zones and as long as new climate in the area has existing analogues any where in the world people in that area can adapt and learn. Thus that is regarded as the best method through which a balance between the rising sea level and man can be maintained and the human species can see traces of survival for their future generation's comfortable and sound life.

Sea level rise is a problem, whose impact will be seen in nearly all the regions, especially coastal regions all over the world. But South Asian Region or the SAS is said to be the most vulnerable one. A vulnerability which is caused by factors like low lying coastal areas. Rapid deforestation, overgrowing rate of population, developing and poor economics and an uneducated population. These factors create all the more space for degradation of the coastal areas.

Apart from the rise of sea level, other impacts of the global climate change, has left the South Asian Region in a very weakened position. The global warming which paves way for acid deposition, deforestation, forest firs, reduction of agricultures, and many other such kinds of hazards.

Global climate change and the rising sea level could also be responsible for the spread of diseases, an increase in their intensity, flooding of coastal areas, droughts, famine etc., a challenge that the South Asian Economics would not be in a position to meet.

Concern about coastal swamps and marshlands comes from their special ecological value and the fact that they are already under stress from human development and pollution. And as these constitute an integral part of the South and South East Asian regions, especially in Bangladesh, Thailand, Indonesia, Sri Lanka, etc. where much of the land is bordered by the sea are in dire need of required steps to prevent the rising sea level, which if risen as predicted could severely hit their social and economic situation.

Special programmes thus have been started in the South Asian region. These programmes would not Just help combat the rise in the sea level but have also taken steps for prevention of the rise.

Now it is widely accepted that the sea level is going to rise. Therefore, island nations and the third world under developed countries, which are situated in mountain belt and are likely to be most affected, should also examine the biological measures they can undertake to mitigate the adverse impact of sea level rise.

Sea level rise (SLR) is a slow phenomenon, and its impact is gradual. It needs long term and precise observations to understand and predict the likely changes that may occur along the coasts of the continents and islands. These impacts of SLR are global in nature but they undergo annual, regional and local variations. They are influenced by the coastal morphology and rainfall, particularly in the countries situated in the monsoon belt on Asia.

In the delta and estuaries of rivers like Mekong in Thailand, Brahmaputra in Bangladesh, Ganges and Godavari in India, Indus in Pakistan, flooding is almost an annual feature and loss of human life and damage to property in coastal areas occur frequently.

On the other hand, the coastline in the tropical countries is a preferred areas, as these regions are more productive and have a moderate climate and encourages fast development of large metropolitan cities as evidenced now in this zone. The coastline is being reclaimed and modified continuously. The coast of development and density of population is very high, therefore, if these areas get submerged due to climate change the loss will be very extensive.

Due to coastal process , hydrography and currents the continental margins, estuaries and islands undergo seasonal and long term fluctuations. The occurrence and disappearance of sandbars is caused due to silt movement. It influences closing and opening of the mouths of the estuaries and the coastal lakes and brings about changes in the hydrology and biological character of the area. Therefore, it is necessary to study the morphology of the coastal margins, estuaries, coastal lakes, back waters and lagoons and understand seasonal and annual variations. The discharge of sewage and organic wastes in estuaries undergo quantitative changes. In many parts, the estuaries get connected to low lying coastal plains, mud flats, and marshes and pollute them, thus large areas become unfit for development. In these areas, production of organic debris and detritus is a continuous process. The disposal and deposition of detritus and silt depends on coastal currents. Such areas are abundant near Bangkok, Calcutta, Bombay and in Shatal Arab in Iraq and are vulnerable to S.L.R.

Flooding changes the pattern of silt deposition, hydrography, rate of flow and mixing of estuarine, waters. Consequently, changes in biology, productivity, diversity of fauna and flora, reproduction and population composition are brought about.

It is necessary to develop a comprehensive, long-term monitoring system using remote sensing, photo interpretation and verification by the ground trucks to study these areas. The ground truck requires continuous data updating, and data comparison, preparation of charts to catalogue annual changes. The study of long term changes will help in combating the increased sea level and its effect on the human beings in south Asia.

The developing countries of South Asia need a special treatment while talking of their prospects in the wake of the rising sea level and the climate change. Their economies need to be developed and only then can they dream of standing on par with the developed world of the North.

As for a sustainable development this is an issue, which naturally must take priority with any nation and so the South Asian nations are no exception, but trying to force their growth to meet the conveniences of the North is quite an unreasonable requirement of the developed world. South Asia cannot be denied its right to grow nor to choose its own pathways to growth. Nor should that right be constrained by new conditions of trade or financial aid imposed in the name of environment, because who would know the meaning of the impact of the rise in sea level, better than south Asia?

A fact that can't be denied is that the South Asian region does need financial support from the developed world, not just to develop, but to develop in the cleanest possible way. The South Asian countries must be enabled to leap frog the environmentally destructive paths pursued in the west since the Industrial Revolution.

To do this, they will need financial support, vastly improved professional and institutional, capacities and the cleanest and latest technologies. Hiding behind the facade, of improving and developing in an environmentally clean way, the developed nations have somehow tried to curb South Asia's more gingerly taken steps in that direction, as was evident during the recently held Earth Summit at Rio de Janeiro, where, the Indian Minister for Environment Mr. Kamal Nath expressed his reservations against the treaty on climate change.

In India's view, the 'safe' and 'clean' technology implied the exclusion of nuclear technology by which the US would like to bring in the controls on national polices through the backdoor as it were. This amounts to using environment to enforce the nuclear non-proliferation treaty. Thus this is not environment that is being talked of but something else altogether. This kind of an international political manipulation is what makes south Asia a part of the world that is disregarded of consideration of problem, which concerns it the most. What the North fails to understand is that if the South Asian Region had funds to develop economies in a sustainable fashion, why would it still risk its whole existence in development. The problem of funding, which again was a major issue at the recently concluded Earth Summit in Rio is not just the North giving the South, monetary and technical help, but also securing its own future on this globe. There is a tendency to regard the costs of protecting the environment as an additional cost we can ill afford. But, much of what we often term 'cost' should really be regarded as investment which will produce a good return in economic as well as environmental terms. So the best way to move on to a sustainable growth would be through the help of the developed North. Environment is an issue for which even the individual of this earth is responsible for. So to allow policy makers to decide and rule the environment keeping in mind their own selfish interests is not part of a healthy scenario.

Because their actions are motivated by a lot of things, and whether their decisions are implemented into is a thing which has to be waited for. As Mr. Maurice Strong in an interview said "Decision makers should not be allowed to hide behind science, in the name of seeking scientific certainty before acting. In science certainty is an impossibility and decision makers are used to making policy decisions under conditions of considerable uncertainty that is what they are good at". Even if they are good at making decisions during conditions of uncertainty.

6 Applied Marine Ecology

Although 71 per cent of the earth's surface is occupied by oceans and only 29 per cent by land, nearly all of the food and raw materials used by man is derived from the latter. This is in spite of the fact that agricultural soil is only a few inches thick and must be cultivated, protected from erosion, and fertilized, while the ocean with its chemical fertility, its photosynthetic production of basic plant food, and its great fisheries is an entirely nautral resource ready to be harvested .

However, fish harvesting in the ocean has definite limits determined by photoplankton primary productivity. It is estimated that maximum sustained yield to man is of the order of 100-150 million metric tons per year and in 1967, 6000,000 tons were being harvested. The ocean is relatively infertile, yet it covers most of the area; much more productive are the coastal areas, particularly the areas of upwelling but these are of comparatively small size.

A large part of the production is taken by predators other than man and part of it must be left to maintain the breeding stock. The possibility of the ocean serving as an abundant future source of food for man is probably exaggerated. Seafood is important mostly as a supplementary source of protein. Man must look to other resource for his main energy requirements. The plankton of the sea represents a possible food supply for man. Its energy value is approximately 4 kcal/g dry weight, and it is more or less palatable.

However the expense involved in filtering and concentrating it is too great to be economical; poisonous species sometime occur, it is not easily digested and assimilated and consequently it has not as yet proved to be a feasible diet. The energy value of the plankton is used by man at the present time primarily as it is transferred into higher links of the food chain. Aside from the fises, the chief marine organisms used as food are the oysters and other mollusks, shrimp, crabs, lobsters, and sea turtles. These are mostly animals of the continental shelf and estaries. There are a number of problems in the use and conservation of marine organisms.

Natural beds of American oysters on the Atlantic coast have nearly all been exhausted through overfising and pollution. Because of heavy erosion of the land, silt deposition has become excessive in most of the bays and estuaries and there is increasing difficulty for oyster spat to find clean hard surfaces on which to set. Surfaces that are loose or covered with silt are not suitable since the spat is very small and easily smothered.

A common practice is to return to suitable areas all shells of mollusks removed or to introduce other suitable hard objects to furnish the necessary substratum for oyster setting. Control of erosion over the watershed would greatly alleviate the problem. The trend is increasing to lease suitable areas of water and to farm oysters in the manner of an agricultural crop. In spite of their position at or near the top of the food chain, the greatest utilizable food resource of the sea is its fishes. Ocean "farming" is largely limited to estuaries and salt mashes.

Yiels in the Orient occasionally reach of even surpass 1000 kg/hectare-yr (880 Ib/acre-yr). Principal species are milkfish, greay mullets, eels, salmon, sea trout, tilapia, oysters, and shrimp. Conversion of marsh into ponds where water level, fertility, and cropping can be controlled has met with varying success in defferent parts of the world, but may be expected to expand as fish harvesting in the ocean itself approaches its yield limit.

Geographic distribution of marine organisms depends on their responses to current, temperature, and physical barriers; their local distribution is affected by waves and tides, type of bottom, salinity, and depth. Major division of the marine bioycle are pelagic (open water) and benthic (bottom). Major communities recognized are the oceanic plankton and nekton biome in the open sea, the balanoid-gastropod-thallophyte biome on rocky shores, the pelecypod-annelid biome on sand and mud bottoms, and the coral reef biome.

Organisms making up the oceanic biome are widely distributed around the world but may be divided into warm-water, arctic, and antarctic faunas. Coral reefs are found only in the tropics. The two biomes on the continental shelf subdivide into warm-water, temperate, arctic, and antarctic faunas and into more restricted regions and subregions. The warm-water faunas are richest in species, especially in the Indo-Malayan and West Indian subregios.

Marine plankton include a greater variety of forms that does fresh-water plankton, although rotifers are nearly absent; cladocerans, less important. They possess various unique mechanisms for flotation. Although abundance varies greatly from place to place and from season to season, plankton is generally much more numerous in neritic coastal waters than in the open sea. Diel movements between the surface at night and greater depths during the day are pronounced.

Mollusks (squids), fishes, birds, and mammals constitute the nekton. The eaxonomic composition of the fish fauna varies with depth. Biolumines ence is exceptionally well developed among deep sea nekton and benthos. Benthos includes a great variety of sessile, creeping, and burrowing forms. It is very abundant in the littoral zone, and decreases in numbers with depth, although individuals are found in the deepest ocean trenches. There is considerable difference in the life-form and species composition of benthos occuring on rocky shores and on sand and muddy ones.

Zonation of species is more prominent on rocky than on depositing shores. Succession and dominance occurs in some situations, but is less important than in terrestrial communities. Coral reefs have many special features, but are now exposed in the Pacific and Indian Oceans to damage by a species of starfish. Food chains in the sea are similar to those in fresh water but the number of links and species composition varies in different communities. Productivity is especially high in regions where upwelling and turbulence bring nutrients from deeper levels up to the surface. The three great biocycles of ocean, fresh-water, and land come into contact around the margins of the seas. The succession from the ocean to land is abrupt and from the ocean to fresh water through estuaries only slightly less so.

Effective use of the ocean resources as food for man is largely limited to the continental shelf, regions of upwelling, estuaries, and coastal marshes. Plankton is abundant and has high productivity in these regions but is likely to be useful to man only as its nutrient material and energy are transferred with considerable loss into the higher food chain links of mollusks, larger arthropods, fishes, and other vertebrates. Costal regions need protection and management for good production and all waters kept clean from pollution of all sorts.

The three great water reserviors of the earth-the seas, the continents and the atmosphere—are not in any sense self contained units but are in constant communication with one another, the relationships take the form of a circulation. Under the influence of solar radiation, water is constantly vaporized and emitted into the atmosphere in the form of water vapour. This process can occur in a great variety of ways.

On the one hand, water evaporates directly from the surface of seas, lakes and rivers, from glaciers and snowfields, or straight out of the ground. On the other hand, in the process of breathing, all living organisms, animals and plants alike, emit both carbon dioxide and water vapour in to the atmosphere. A third process is the emission of water vapour by the combustion of organic materials such as wood, coal and oil, which are to be found all over the earth. Easily the highest proportion of vaporized water emannates directly from the oceans, however. The water suspended in the atmosphere in the form of water vapour is of the greatest importance to the climatic conditions on earth.It conditions the compositon of the air masses, exerts an influence on the energy conditions of the atmosphere, and substantially governs and sustains the water circulation.

The water content in the atmosphere is measured in terms of the relative humidity of the air: Whereas some portions of the vaporized water masses remain permanently suspended in the atmosphere in the form of moisture, others are condensed back by the cooling down of ascending airstreams and emerge in the form of clouds, fog, rain or dew.

If the water cools still further, snow or hailstones may build up in the clouds. When the clouds get too heavy, precipitation begins and the moisture returns to the earth in the form of rain, snow or hail. If the precipitaiton goes straight into the sea or into lakes having no outlet, the water circulation will be complete. If it falls on land, some of it will accumulate by surface drainage in brooks, streams and rivers and so return to the ocean.

Other forms of precipitation seep into the earth and replenish the groundwater, or if retained at an upper ground level as trap water will evaporate from there. The trap water is also available to sustain plant life, which absorbs it through its roots and returns it to the atmosphere as water vapour by respiration and transpiration. This circulation, like all others on the earth, cannot occur in the absence of some motive force. The entire cycle is maintained by the regular inflow of solor energy. Of the total amount of energy absorbed by the earth, about one-third is utilized in maintaining the water circulation.

The temperature, humidity and movement (wind) in the atmosphere together determine the level of evaporation. Although the toal volume of atmospheric moisture is very small in comparision with the total hydrosphere, owing to the circulation of the atmosphere an enormous amount of water movement takes place in the course of a year. To produce the annual precipitation on the earth of about 470,000 cubic kilometers, the wate in the atmosphere about 12,300 cubic kilometers) must be replaced 35 to 40 times a year.

Sunlight penetrates the sea to a depth of about 1000 m (3000 ft). Animals living in this twilight zone consequently have very big eyes and exceptionally wide pupils. Like many nocturnal animals on dry land, deep-sea animals have all the sensory cells of their retinas protruding on rods. These rods, which serve to pick up what little light there is and convert it into nerve impulses, are unusually long and large. A nerve cell from the optic nerve is linked simultaneously to several rods. Under this arrangement an extremely high sensitivity to light is achieved. At a depth of below 1000 m (3000 ft. or so, at which almost total darkness prevails, the fish have either no eyes at all or only very small eyes of limited capacity. It is now known that such deep sea fish find their way in the dark with the aid of sensory organs with which they are able to detect vibrations in the water.

Many deep sea fish have very long organs of touch of a most remarkable kind: tentacles of great length with feeler-type floating ends and extensive tail ends. Many fish have long feelers furnished with claws and sharp points, which also serve to locate and trap their prey. Certain sea organisms are able to illumine their paths of their own accord or to radiate more or less intensive light (bioluminescence). In many cases a glow is cast by certain cells which are surrounded by reflecting cell groups. Other animals can emit luminous substances into the surrounding water which provide light away from the animal concerned for some time and probably serve to camouflage its activities.

In particular cases a symbiosis is established between luminous bacteria and fish whereby the luminous bacteria are in effect kept in a tissue culture by the larger animal. Some varieties emit projector like light beams that probably help to locate the prey. Other deep sea fish have luminous bait at the end of long extensions, which take various forms and colours. In some varieties the luminous bait even hangs from an enormous jaw. In other cases light organs may help the male and female of a particular species to locate one another on the seabed for mating purposes. Many sea organisms can switch their light beams on and off, as it were , so as to confuse their opponents. Many deep-sea fish have enormous teeth and jaws. And their jaws are often extremely flexible. This points to the fact that their nutriment may be found only at rare intervals, and it may then be very bulky and have to quell the animal's hunger for long periods. A number of aspects regarding deep-sea fish still remain to be clarified, however. We often have to depend on chance discoveries, and animals that are brought to the surface are apt to die quickly, so that intensive investigations are not possible.

Like the mainland and inland waters, the oceans of the world are also subject to seasonal temperature variations. In the sea, however, these changes are more limited. This is due to the water circulation that takes place constantly within the oceans, coupled with the extremely high specific heat of salt water and the capacity that goes with it for storing energy. Except in the shallow shelf that surrounds the continents, the temperatures in the seas are consequently very uniform.

In low geographical latitudes on each side of the equator, temperatures of up to 30°C (86°F) are recorded in the surface waters, while in shallow and enclosed waters they may rise in coastal regions to as high as 50°C (122°F). Except in extreme cases the temperature variations between the coldest and warmest areas are in the 30°-35°C (54°-63°F) range. In polar regions and in equatorial waters the temperatures over the year remain remarkably uniform. As for deep-sea water, a constant water temperature can be assumed in all the oceans. The lowest temperatures are to be found in polar sea regions, where they can come close to the freezing point of seawater (-1.9°C, or 28°F).

Whereas in the lower latitudes seawater absorbs warmth from the atmosphere, at higher latitudes it emits warmth into it. The cooling of the water surface caused in this way leads to the formation of convection currents, which results in the intermingling of the water masses and the equalization of temperature between surface and deep water. This phenomenon arises from the fact that sea water (salt content 35%) attains its maximum density at –3.5°C, or 25.7°F (compared with 4°C, or 39°F, for pure water). In this respect sea water behaves quite differently from freshwater. It freezes on cooling before it reaches its maximum density. From this it follows that no stable temperature stratification takes place in winter or in arctic regions, as in the case of freshwater, but that the intermingling in the sea is not interrupted.

The heat accumulated in sea water is emitted into the atmosphere as the water cools, not as in freshwater where below a shallow surface layer the water maintains a constant temperature of 4°C (39°F). It is from this property that the sea acquires its importance in moderating the climate. Because of its lower specific gravity, at lower latitudes the warm surface water cannot promote convection currents. This gives rise to considerable drops in temperature between the surface and deep water levels and to the development of so called boundary surfaces between the two layers, which occur at depths of 100-500 meters (328-1640 feet).

In temperate latitudes a seasonal change takes place in the formation of convection currents and boundary surfaces. During the summer months, when the surface water is warmed by increased solar radiation, the boundary surfaces form at a depth of 15 – 40 meters (49 – 131 feet). In the autumn, when the water is gradually cooling off, the boundary surfaces disappear and convection currents build up.

Sea water is a complex solution of various salts, trace elements and gases. There is a measure of equilibrium between the input of soluble matter by the rivers, elutriation (or removal of suspended particles) from the atmosphere by rain, and the loss by precipitation and absorption of nutrients. The bulk of the soluble matter is washed into the sea through the weathering and erosion of rocks from the solid crust of the earth. Much of this matter is soon precipitated to form new rock deposits on the seabed, which become stratified over geological periods and can form new mountains again as a result of pressure movements in the earth's crust. In shorter periods, variations can be caused in the composition of seawater by biological processes (occasioned by the temperature, for instance) and today in particular by the deposit of waste and toxic substances in the sea. In general, the quantity of inorganic material dissolved in sea water amounts to 35 grams per kilogram of water. This represents a solution of 3.5%.

To come down to details, sea water contains the following elements (average concentration):

Substance	*Concentration (milligrams per litre)*
Sodium	10,600
Chlorine	18,990
Magnesium	1,350
Sulfur	885
Calcium	400
Potassium	390
Carbon (As carbonates and Carbon dioxide)	23-28
Strontium	8.1
Bromine	65
Boron	4.6
Silicon	0.02 - 4
Fluorine	1.5
Nitrogen	0.02 - 0.8

Rubidium	0.12
Aluminium	0.01
Lithium	0.18
Barium	0.05
Iodine	0.05
Phosphorus	0.0001 - 0.1
Arsenic	0.003
Iron	0.002 - 0.02
Manganese	0.001 - 0.01
Copper	0.001 - 0.01
Zinc	0.005 -0.01
Lead	0.00003
Selenium	0.00009
Cesium	0.0004
Uranium	0.003
Silver	0.0003
Thorium	0.00005

Most of these elements are present in sea water in the form of salts, and a smaller portion in the form of dissolved gases. The most important of these are oxygen and carbon dioxide. The oxygen content fluctuates between 0 and 8.5 milliliters per liter. The high concentrations occur near the surface of the water, where the oxygen is in equilibrium with the oxygen in the atmosphere.

A second source of the oxygen dissolved in the sea comes from the photosynthesis of phytoplankton, the drifting oceanic plant life. At great depths it may happen that (owing to the intense activity of bacteria and animals that absorb oxygen) the oxygen content of the sea water falls to zero. In many places, however, owing to the downward trend of the convection currents, a large quantity of dissolved oxygen from the surface water is carried to considerable depths in the sea.

An important part is played by carbon dioxide gas, which is present in considerable quantities in sea water. Since the basic ions of sodium, potassium and calcium are also present in large quantities in sea water, this facilitates the solution of relatively high percentage of carbon dioxide. Carbon dioxide is one of the basic substances for photosynthesis (see part six).

The plants living in the sea (especially plankton) take the carbon dioxide they need for photosynthesis directly out of the surrounding water. The second important effect of carbon dioxide in sea water is its function as a buffer (dissolved matter which keeps the pH value of the solution practically constant over wide areas). Normally the pH value of sea water lies between 7.5 and 8.4. As carbon dioxide and its associated marine chemicals are in equilibrium, the pH value of sea water is approximately constant as regards the input of both acids and bases.

The other gases from the atmosphere are dissolved in sea water in much lower concentrations than oxygen and carbon dioxide. The concentrations are as follows:

Substance	*Concentration (milligrams per liter)*
Nitrogen	between 8.4 and 15
Helium	0.000005
Neon	0.0001
Argon	0.06
Krypton	0.0003
Xenon	0.00005

A particular phenomenon is the enrichment of sea water by trace elements that occur in considerable concentrations in living organisms. An example is provided by the ascidian (a tunicate), a small, mostly clinging animal that increases the element vanadium in sea water up to 50,000 times its normal concentration.

It is known today that many other trace elements such as iodine, arsenic, nickel, zinc, titanium, chromium and strontium are also very much increased in the water by decaying organisms. Certain fish are known to contribute chromium, nickel, silver, tin and zinc. Similarly, some of the substances produced by humans get into the sea as waste substances. As such a build up of substances in the food web takes some time, the damage caused by them is often not noted until too late. The increase of chlorinated hydrocarbons, for instance (especially pesticides such as DDT), reaches its maximum concentration in fish only after a period of 11 years from its production. Also of great importance in its effects on life in the sea is the content of ammonium compounds, nitrates and phosphates. As these are absorbed by plants to build up their bodies, their content in sea water varies considerably according to the presence of plant life. The following concentrations in sea water have been established:

Substance	*Concentration (milligrams per liter)*
Ammonia	0.4-5.0
Nitrate	1-600
Nitrous oxide (particularly NO_2)	0-15
Organically combined nitrogen	30 - 200
Phosphates	1-100
Organically combined phosphorus	1-30

The oceans cover nearly 71% of the earth's surface, or about 361 million square kilometers. This enormous volume of water is not spread evenly over the earth's surface, but varies sharply in area and depth. Three extensive oceans separate the main land masses of the American, Eurasian and African continents from one another. With an area of 180 million square kilometers the Pacific Ocean accounts for nearly half the total sea surface of the world.

7 Marine Pollution

Nearly 71% of the earth's surface is covered by oceans, which together comprise a total of approximately 1.37 x 10^{39} litres, weighing about 1.40 x 10^{36} metric or long tons. This is the tremendous amount of water and consequently for many years the oceans are regarded an ideal place to dump all of man's wastes. For example, the rivers of the world discharge close to 2 x 10^{14} metric tons of water into the ocean every year but this is much smaller than the water already into the ocean. Carried along with the water each year are 4.5 x 10^{11} metric tons of dissolved materials. Although this is a large amount of material compared to that already present it is literally a drop in the bucket.

Unfortunately, the ocean does not receive all of its dissolved material input form the world's rivers. Some comes from the atmosphere as it is washed by falling rain. Most of the pollutants placed into the atmosphere and up in the oceans either directly or indirectly form precipitation. This also is a sizable amount of material. It is to be noted that this is larger than the amount of sediment plus the amount of dissolved materials carried to the oceans by rivers. The more important reason for being concerned about marine pollution stems form the fact that materials put into the oceans by man's activities is not evenly spread throughout the oceanic volume, pollution is generated where man lives and works and therefore, finds its way for the most part into the oceanic areas that are closest to their activities. Consequently, the dilution factor is very small and the concentrations of these pollutants build up in coastal regions. They are increased to a much higher level than would be the case if the material were distributed throughout the entire oceanic volume. Coastal are dumping grounds area having a much higher pollution concentration not only because the material is being put into these relatively shallow areas much more rapidly than it is being carried away by natural nations but also because the normal structure of the oceans tends to prevent the mixing of these inputs with the rest of the oceanic volume.

The oceans are not dead from pollution, not are they insensitive to the effects of pollutants. As long as we find it necessary to use the sea for a world dump, and there is no reason to believe we will ever stop, we must learn as much as we can about oceanic processes so that the assimilative capacity of the oceans is never exceeded. Strictly speaking marine pollution may also be of natural origin. For example, it may be caused by underwater gas and oil eruptions, or by the intensive growth and subsequent death of certain micro organisms, which are responsible for the familiar disastrous red tides. However, such processes and effects are not regarded as 'pollution,' even though they are often studied by methods similar or those used for investigating pollution effects caused by man.

Ecological studies on marine pollution fall into three principal categories:

(1) Biogeochemistry of pollutants.

(2) Marine ezcotoxicology.

(3) Biological principles of anti-pollution measures.

The biogeochemistry of toxicants includes the investigation of the sources of pollutants, the pathways along which they enter the marine environment, patterns of accumulation in the biotic and abiotic components of ecosystems, mechanisms and rates of migration of pollutants, their transformations and other processes which determine the fate of toxicants in the sea.

An important, though not decisive stage in such investigations is the determination of the content of toxicants in marine organisms and communities; this is a difficult analytical problem because the chemical composition of matter is complex, while the concentration levels of pollutants are low. The methods of chemical oceanography are used in order to obtain and analyses samples, the results being interpreted in a biogeochemical sense.

The second type of research work carried out on this problem, which is best described as 'marine ecotoxicology,' involves the study of biological effects and consequences of human interference with the composition of the marine environment.

Finding answer to such questions, is not by itself tantamount to solving the problem of pollution, but nevertheless yields an objective evaluation of presentday ecological anomalies in the world ocean and prediction of those likely to occur in the future. These investigations provide a scientific foundation on which to base national or international efforts to preserve the seas from pollution. Solutions of the immediate practical problems concerning pollution of the sea must be based on the results of closely interlinked ecotoxicological and biogeochemical studies.

Marine Pollutants

In Table 1 a list of common pollutants associated with the marine environment have been given. These are not listed in any particular order of importance because it is found that in certain areas certain pollutants are more worrisome than others, while the opposite condition may be true in other areas. This difference in importance comes from local conditions like population pressure, flushing, and climate. Theoretical calculations of the concentrations of the principal toxicants in epipelagic waters of the World Ocean indicate that considerable pollution (lead pollution, in particular) of surface waters by atmospheric pollutants may have taken place. This is true of inland seas. The global pollution zone of the marine environment may be characterised as follows: highest pollution levels in the euphotic layer; high pollution levels in the neritic zone and inland seas; latitudinal distribution of toxic pollutants; a mosaiç distribution pattern of toxicant concentrations in water; loalisation of toxicants in the hyponeuston and benthos biotope; and a coincidence of the maximum pollution zones with zones of high biomass and high productivity.

Table 1

Pollutants Associated with the Marine Environment

Pathogens	Toxic organics
Sediments	Petroleum
Solid wastes	Nutrients
Heat	Radioactive materials
Fresh water	Oxygen demand materials
Brine	Acids and bases
Toxic inorganics	Aesthetically displeasing materials

Before we discuss the marine pollutants one by one, we shall give some general conclusions about these:

(1) The rate of biological accumulation and migration of micro-amounts of metals and radionuclides in the World Ocean is directly correlated with their physichemical properties an with the specific surface area of aquatic organisms.

(2) Concentrations of metals in sea water and in plankton tend to decrease with increasing atomic number of the metal.

(3) The relative content of artificial radionuclides in oceanic seston does not exceed 1 per cent of their total content in water; the corresponding figure for heavy metals and transition is 1 - 10 per cent, while in inland seas it may be as high as 25 per cent.

(4) The forms of microcomponents in sea water are largely determined by the suspersaturation of surface water by calcium carbonate, since the suspended and the colloidal forms of this compound are strong sorbents.

(5) Self-purification of surface waters by bio-sedimentation of impuprities is mostly due to their accumulation in aquatic organisms at low trophic levels. If the accumulation factor in seston is 1000-10000, the rate of vertical biogeochemical migration of a given impurity is of more or less the same order of magnitude as the rate of its hydrological transport (for biomasses and productivities considered as average for the World Ocean). This is true of many radionuclides (^{144}Ce, ^{95}Zr, ^{106}Ru, ^{65}Zn, ^{54}Mn, etc.), of heavy metals and transition metals present in microamounts (mercury, leads, cadmium, iron, zinc, etc.) and chloro-organic compounds (DDT, PCB, aldrin, lindane, etc.).

(6) The rate of input of petroleum and chloro-organic toxicants into the World Ocean is faster than their

rate of decomposition under natural conditions. The degradation processes are particularly slow at great depths, and the latter must be considered as a depot of toxic substances and their metabolites.

(7) A typical feature of the geochemical behaviour of petroleum and chlorinated hydrocarbons in the World Ocean is that their distributions in biotic and abiotic components of the same ecosystem are altogether different.

(8) The contemporary radio-ecological status of the World Ocean is distinguished by the relatively stable contents of ^{90}Sr and ^{137}Cs in surface waters and in aquatic organisms. Another typical feature is the gradual increase of radionuclide contents in the ecosystems of inland seas.

(9) The concentration levels of artificial radionuclides (^{90}Sr and ^{137}Cs), heavy metals (mercury, lead and cadmium) and chloro-organic compounds (DDT, polychlorinated biphenyl, etc.) in commercially important marine products vary with the degree of pollution of marine biotopes, taxonomic identity of the organisms, and their ecological and physiological characteristics (trophic level in the community, diet, age, zinc, etc.).

The most variable concentrations in industrial species are typical of substances with cumulative properties, DDT and DDT metabolities, polychlorinated biphenyls and other chloro-organic compounds, and methyl mercury.

(10) There is a generally valid relationship (a direct corelation in the case of radionuclides) between the concentration levels of toxicants in commercial animals and the pollution level of the marine environment as shown by the increasing sequence of

pollutant concentrations: pelagic ocean waters, nertitic zone, inland seas, freshwater basins.

(11) Typical accumulation factors of ^{90}Sr and ^{137}Cs in commercial species are of the order 10^1-10^2 for heavy metals 10^2-10^3, and for chloro-organic compounds 10^3-10^4. The food chain effect - i.e., the dependence of the toxicant concentration in a given organism on the trophic level of the organism in the community- is definitely present only in the cases of mercury, polychlorinated biphenyls, DDT and other chloro-organic compounds. The contents of mercury and of chloro-organic compounds in fish show a regular increase with the size (weight, age) of the animal.

(12) The selective, irreversible accumulation of high concentration of mercury in pelagic predatory fish and marine mammals is a natural process, and is not usually due to human factors.

(13) The distribution of toxicants in the organs and tissues of commercial fish species is typically non-homogeneous. ^{90}Strontinum, lead and cadmium become preferentially concentrated in ligaments and bones; chloro-organic substances and methyl mercury concentrate in the liver and in the lipid fractions; ^{137}caesium concentrates in muscles. Chloro-organic compounds concentrate in factory tissues.

(14) Global concentration levels of radioactive and chemical toxicants in commercial marine products are usually much lower than the national and international norms for maximum permissible concentrations, and thus present no danger to human health. However, in view of the possibility of local pollution and human interference with the natural background of micro-components in individual regions, concentrations of toxicants in

commercial species and marine products must continue to be constantly monitored, especially for heavy metals and chloro-organic compounds.

(15) Generally speaking, monocellular algal my react to the presence of toxic components in the aqueous medium in two different ways:

(a) By inhibition of vital processes and production.

(b) By stimulation of cell development and cell division, which is usually followed by impaired vital activities.

The development of micro-algae is relatively independent of the presence of the toxicant in the medium.

(16) The inhibitory effect of the pollutants increases in the following sequence: oil, detergents, metals, (arsenic, lead cadmium, copper, mercury, methyl mercury), chloro-organic compounds. The most highly toxic compounds are DDT, polychlorinated biphenyls, mercury and methyl mercury, which inhibit photosynthesis in long-term experiments at conentration as low as 1-10 *m*g/1. Arsenic, when present as arsenate ion in concentrations of up to 1-10 mg/1, had no effect on the rate of division of algae cells.

(17) Low concentrations of toxicants (0.1-10 *m*g/1) acting for a short period of time tend to stimulate photosynthesis and the rate of cell division in algae. This may be caused by the overall activation of vital processes by small doses of poisons or by an alteration of the metabolic interrelationships with the cells of the microflora. In the case of petroleum products such a stimulation may be due to the presence of biologically active substances in petroleum.

(18) Toxic (algicidal) and threshold concentration ranges are determined for various types and species of monocellular algae. Out of the several combinations of the effects of several factors studied, synergistic effects were observed for the combinations of petroleum with DDT, and combinations of certain metals with a detergent. A combined effect of several metals may be additive, or else may be manifested by a dominant inhibitory effect of the most toxic one (mercury, copper).

(19) The varying response of different monocultured species to toxic impurities in short-term experiments (less than one day) are probably due to the differing rate of absorption and penetration of toxicants into the cells. In long-term experiments the specific differences are usualy much smaller and depend on molecular, cellular and tissue level mechanisms. A general characteristic is the better resistanace of freshwater protocol algae to most pollutants, as compared with cultured marine phytoplankton forms.

(20) Of the species studied, Gyrodinium fission in highly sensitive to mercury, the response being irreversible; the same applies to diatomous algae acted upon by a mixture of DDT and petroleum products.

(21) The inhibition of photosynthesis is a more sensitive reactions of monocellur algae to toxicity than is the decrease in the rate of cell division.

(22) The general picture of the effects and relative toxicities of the most common toxicants on natural communities of marine phytoplankton does not significantly differ from that obtained for cultures of single species.

(23) In a number of cases the intensity of photosynthesis of natural phytoplankton communities decreases in the presence of lower toxicant concentrations than in experiments carried out on monocultures. The higher sensitivity of natural phytoplakton to contaminants is probably due to interference with the interspecies relationships in the community.

The considerable geographic, seasonal and other fluctuations in the reaction of natural phytoplankton to the presence of toxicants are caused not only by the environmental factors, but also by the special features of its specific composition. This composition undergoes significant changes if the action of the contaminants on the phytoplankton is sufficiently prolonged.

Pathogens

The pathogenic materials are the living organisms that can produce sickness or biological unbalance in either plants or animals within the ocean itself or in humans who either come into contact with oceanic waters or eat the organisms caught in the water. These include a wide variety of bacteria, protozoa, viruses and fungi. The most common of these are normally found in sewage. However, other pathogens may occur in non-waste disposal areas where environmental conditions are such that the proper conditions exist for growth and reproduction.

Sediments

Sediments are always present to some extent in the marine environment. These sediments have a marked effect on plant growth because they block out a large portion of the light normally reaching greater depths and therefore decrease photosynthetic activity. In some case rooted plants get completely destroyed by this process.

As sediments are deposited on the bottom they will cover up bottom dwelling organisms (benthos) such as oysters and may even smother them under extreme conditions. The deposited sediments tend to make the water shallower so that, in regions traversed by ships, periodic dredging is needed.

Solid Waste

The disposal of solid waste has been a critical urban problem because areas suitable for the dumping of these valuminous materials are becoming scarcer. Due to this reason the ocean has been used as a dumping ground for solid waste. The solid waste most frequently marine dumped is sludge material left over as a by product from domesic sewage treatment. However, the unused products of industry and the used-up products of society are also put into the sea. If the product discharged at sea contains materials which may be leached into the oceanic environment, a serious problem could arise especially when the materials are toxic. However, when the solid materials are inert, a little forethought may result in some benefit to the marine environment.

Excess Heat

Excess heat, if added to the marine environment, alters ambient conditions, and these changes may be detrimental to the organisms present. The amount of heat that is detrimental and the extent of the degradation is determined by a number of factors. The primary source of this heat has been of course, from electrical generating plants, whether they be fossil fueled or nuclear powdered.

Fresh Water and Brine

Although fresh water may be in great demand ashore, too much of it in the ocean obviously will produce a marked environmental change within a small area. This change may occur due to poorly designed storm drainage systems, wear diversion networks associated with dams, and effluent of some industrial processes. Excessive fresh water usually is not a critical problem. The introduction of brine into the marine environment is similar in its effect to that of fresh water except in the opposite direction. The organisms acclimated to a particular salinity now find themselves in a more saline environment which could make permanent damage. Brine is a by product of desalination plants.

Toxics

Toxic inorganics material are commonly used in industry, many of them are relatively harmless; however, in larger quantities they may be quite destructive. There are perhaps, 35 to 40 commonly used toxic inorganics and these should be very closely controlled by the user. The toxic organics have been the most disturbing of the modern day chemicals commonly discharged either purposefully or accidently into the marine environment. These include the biocides such as fungicdes, herbicides, insecticides, rodenticides, and also the additional organics including halogenated hydrocarbons, petroleums and industrial chemicals.

The most did turbing toxic organics have been the pesticides such as DDT and ketone which have the unfortunate, characteristic of being more soluble in oil than in water so that they tend to collect within the fatty tissues of marine organisms. They have been also very stable compounds which do not deteriorate very easily over a long period of time. These find their way into the ocean both as manufacturing effluent and as runoff after utilization.

Petroleum

Although petroluem may be classed as a toxic organic, it is a naturally occuring material and is biodegradeable, given enough time. Its effects are not too well known. Petroleum enter into the marine environment due to accidents such as tanker damage or transfer loss, natural seepage, off shore production losses, losses associated with refineries, from runoff originating as drippings or disposal of used automobile lubricants, and unburned hydrocarbons emitted into the atmosphere as internal combustion exhaust.

Nutrients

Nutrients are those chemicals which are required by plants. The activities of man have added to the total nutrient load of almost all coastal areas. When nutrient levels becomes out of hand, plants grow unchecked so that decaying plants exist in such great numbers that the oxygen supply becomes rapidly depleted. These nutrients are present in domestic sewage effluents, agricultural runoff, and the little understood but apparently important non-point source runoff from urban areas.

Radioactivity

Radioactive material have been not only discharged to the marine environment by nuclear power plants, nuclear power plant fuel production and reprocessing plants, and uranium activities of all sorts, but also result from more common activities of burning of coal. When coal is burned it gives out radioactive particles to the atmosphere which are then washed into the sea at a greater rate than any known nuclear power plant at this time. Other sources of radioactivity include the natural back-ground, weapons testing, mine drainage, accidental spillage, and a few isolated industries. One of the partially unsolved problems associated with the use of nuclear energy for electrical power has been the long term storage or disposal of spent fuels. This storage must be in an area such there have been no pathways back to man, and the deep oceans have been suggested as meeting this criterion.

Oxygen Demand

Oxygen demand materials have been those which need oxygen for degeneration and therefore steal oxygen which would normally be utilized by marine animals. Thus if too many oxygen demand materials are kept in the marine environment, the animal population will be markedly decreased due to the lack of oxygen. Sewage sludge and any other organic waste material, even that resulting from excessive plant growth due to an over supply of nutrients have examples of common oxygen demand materials.

Acids and Bases

The discharge of acid and bases to the marine environment is quite disturbing to the natural ecological balance of the system. The normal pH of oceanic water has been somewhere around 8.0, slightly basic. This is maintained by the carbonate system. If a large amount of acid or base is introduced into the system, the carbonate reactions will be offset and an important element of the environment will get affected. In addition, there are large synergistic effects associated with pH. Most toxic materials, for example, increase their toxicity under conditions of low pH. The sources of acidic or basic material have been primarily industrial with some of this material reaching the marine environment from accidental discharges and the rupturing of tankers.

Aesthetic Considerations

Aesthetically, displeasing material include all the stuff one finds in the ocean which is unpleasant to look at or to smell. Tar balls, floatables, gas producing materials, and colouring agents are some examples, of pollutants offending the senses. Although in some cases these materials do not pose any real threat to the ecology of an area, when the area is being used for recreation, the quality of the surroundings becomes some what important.

Although the major pollutants are delineated, the management of marine pollution is quite complex. The reason for this is that the marine environment is used by man in many different ways and this mutli-use is having a number of ramifications. In the first place each oceanic activity pursued by man may produce a whole range of pollutants. A moving ship, for example, may not only discharge pollutants in the form of waste products from the people living aboard, oil from pumping its bilges, and heat from the discharge of its condensers, it may also need deepening of a harbour channel. This may generate a salinity change in addition to sedimentation problems associated with the maintenance dredging required. Each pollutant may come from many different users. The nutrients which primarily appear to come from the effluent of sewage disposal plants also may come from storm drainage of urban areas in the form of waste products of domestic animals, along with runoff from agricultural areas that have been fertilized by either commercial fertilizers or farm animals.

An example of the effect of one pollutant or another has been the attraction of heavy metals to suspended sediments. Most of the heavy metals in coastal waters are swept out of the water by sinking sedimentary particles and end up in the bottom deposits. If the heavy metals then enter their way into bottom dwelling organisms which, in turn, find their way into a food chain pathway to man, then this obviously is not desirable. Another example of this sysnergistic effect of two or more pollutants has been the effect of oil on pesticides, combined with sediments. The pesticides are more soluble in oil than they are in water so that if there is oil present in oceanic water, it will tend to concentrate any pesticides that also may be present. Oil is also having an affinity for sediments and as it collects on the suspended particles, their weight gets increased to the point where they sink to the bottom. Portions of the inshore marine environment thus are covered with sediments very rich in both oil and pesticides. The introduction of foreign material into the marine environment has not always been detrimental to man or to the environment either. In many areas the increased erosion produced by construction activities in large areas of siltation where these eroded sediments have been deposited in the shoreline areas. After stablization, populations of marsh grass have been seen to spring up in many of these newly created areas. In the main there is no planning for this type of rehabilitation, but recent experimental work reveals that production of marshlands may very well be a common tool of the estuarine planner of the future.

Now that the pollutants are delineated and described very briefly, we will concern ourselves with the activities involved in the production of these pollutants in an attempt to pinpoint the source of the problem. These activities are outlined in Table 2.

Table 2

Pollution Producing Activities

Marine Commerce
Industry
Electrical power generation
Sewage treatment
Other non-industrial
Recreation
Construction

Marine Commerce

Marine commerce has been deeply involved with ships, and ships require a deep enough water environment so that they can pass to from on their appointed rounds. This usually requires continuous dredging. Dredging is an expensive proposition requiring large expenditures of money. Once the sediments are extracted from the bottom and the channel is cut to the desired depth, the problem becomes one of disposing of the dredge spoils, the material dredged from the bottom. Various solutions have been suggested but the problem is still one that has not been adequately solved. Another aspect of the ship problem has been oil. Whether these ships has been transporting oil or using it for fuel and lubrication, there always seems to be some lost to the sea. Most harbours are having large storage areas open to the atmosphere so that a certain amount of leaching into the marine environment is inevitable. Lastly, there has been a whole family of pollutants associated with any ship. A ship is just a little city afloat and any problems found in urban areas will be found on ships also.

Industry

Whenever one mentions pollution to the average individual the first source of pollution that pops into mind has been industry. Most industries require the use of water. This water has either heat or some chemical added to it before being discharged into the environment. These chemicals may be anything from toxic substances to harmless coloring agents. Electrical power generation is responsible for waste heat and certainly this is one of the by products of producing electrical power whether it be from the burning of fossil or nuclear fuels.

An additional problem involved with heat exchangers has been the fact that many of these are made from copper, a toxic material which is used in anti-fouling paints on boats for many years. Sometimes these heat exchangers are allowed to remain in an unused condition for extended periods of time with water sitting in the heat exchanger tubes. Some copper leaches into this water form the tubes and when the plant gets turned on again this copper-rich water is pumped into the environment where it may bring about some consternation among the local marine organisms.

Large amounts of water bumped at high speeds from a coastal regions are able to trap or entrain small of sometimes even large organisms. These creatures may suffer mechanical damage as they go through the system, or they may get exposed to large amounts of heat for a short period of time. Radioactive wastes are considered as one possible disadvantage of nuclear power plants, but the amount of radioactive waste actually transferred by cooling water to the surrounding environment has been measured as extremely low. The spent fuel problem appears to be of more concern at this time.

The pollution problems associated with mineral extraction could not be documented due to the small opportunity for data taking, but it is expected they will be similar to marine construction. At this time the major oceanic mining effort has been directed toward the extraction of sand and graved in the near shore areas. Within the imminent future it is expected that a connected effort will be made to extract manganese nodules from the ocean floor, perhaps at depths as great as 300 metres. This mining effort will be expected to have some effect on the environment but just how much has been not quite clear.

The same can be said for the increase in activity connected with offshore oil drilling although it appears that there is less petroleum lost into the environment when the oil is obtained from an offshore well and transported to the continent by pipeline than when it is obtained from a land well and transported by means of tanker to its enventual user point. The spillage resulting from the use of tankers has been greater than that resulting from the use of pipe line.

Sewage Treatment Systems

The process treating human waste products has been quite successful in removing suggested materials and pathogenes, but it has been extremely difficult to extract all of the nitrogen and phosphorous compounds. Consequently, sewage treatment facility effluent is having reasonably high amounts of these nutrients which push plants toward uncontrolled growth. In addition, when the facility gets overloaded, as during a heavy rain, it is often completely bypassed, discharging raw sewage directly into the water. This needs a great deal of oxygen for decomposition so that the biological oxygen demand is increased. One aspect of sewage treatment neglected up to now, primarily because the costs have been so overwhelming, has been the treatment of storm runoff.

Unfortunately the solution is not as straight-forward as simply increasing the capacity of existing plants to handle the heavy loads associated with storms, or even to add treatment to storm drainage systems when they exist separately. Much of the urban and rural runoff does not go through conventional pipe systems, so additional systems would have to be built if all runoff has to be treated. Storm drainage washes urban streets clean.

All material deposited on the streets, including crankease drippings, rubber worm from auto tires, and bird, dog, and cat droppings are washed to sea. Modern studies have indicated that this 1 runoff has been a very significant source of coliforms and petroleum, to name just two pollutants of some concern to man. How to handle these nonpoint sources has been a task presently facing marine environment managers and one that appears to have no inexpensive or simple resolution. With most large urban areas disposal of the solid residue from sewage treatment plants has been a solid waste disposal problem of the first magnitude.

Ocean dumping in the nearshore coastal regime is utilized for many years and, to a certain extent, will probably continue to be. Either the sludge must get treated more thoroughly before it is dumped or new deep sea dumping means must be found. If new treatment or disposal techniques are not followed, there may occur major changes in the coastal environment.

Agricultural Sources

The largest single class of non-industrial polluters are farms. Agricultural pollutants have been commercial fertilizers, animal wastes, pesticides, herbicides, and sediments. In many cases the agricultural contribution of all these pollutants has been greater than any other single source. The pollution can be stopped, and there have been cases in the past where particularly bad problems have been solved.

Recreational Activities

Usually when we think of marine recreation we think of a non-polluted area since recreation is essentially limited by pollution. In shallow water motor boats stir up the sediment on the bottom, a condition which often persists for long periods of time. This increase in turbidity markedly influences the natural plant growth and probably has something to do with the decrease in the numbers of rooted plants reported in the last few years. The aspect of motor boating that has many investigators most concerned has been the discharge of petroleum products into the environment. There are some figures which indicate that recreational boats add as much petroluem and petroleum waste products to the environment as all other source combined.

Construction

The last source of pollutants listed in Table 2 has been construction. By construction has been meant both that done at sea such as offshore oil rigs, and than done on land cotiguous with the sea, such as port facilities. Implied here has been both the actual construction of the structure and the associated activities that go along with construction. There must be all sorts of support equipment, but most important, there must be support people.

As in general, the amount of pollution gets directly related to numbers of people, any large construction effort will bring with it the usual people waste products. The effect of a new physical structure where one had not previously been too well known, but one thing is sure—the environment will never be the same as it was before. Some of these changes might be beneficial, but no matter how beneficial there have been always going to be some members of society who react to any change by claiming a degradation of the environment. Any marine construction project must get judged on the basis of the benefits to be derived by society as a whole rather than solely to those accruing to some small group.

Effects of Marine Pollution

In this article, we shall attempt to get some idea as to what pollution does and why we consider it to be so undesirable. There have been three basic types of pollutants; the pathogenic, the aesthetic, and the ecomorphic. Pathogenic pollutants have been those which cause disease. This disease may be fatal if the pollutant is a fetal poison.

Aesthetic pollutants have been those pollutants causing a change in the environment displeasing to the eye, ear, or nose of man. Ecomorphic pollutants, on the other had, have been those pollutants which produce a change in the physical characteristics of the environment in such a way that there may be drastic changes in the structure or composition of the biosphere. Obviously, these three types of pollutants have been not of equal importance. Pathogenic pollutants have been certainly much more serious than the other two; however, ecomorphic pollutants are often an indicator of more serious types of pollution to follow. Furthermore, since pollution has been man-produced and the effects have been suffered by man, aesthetic problems are certainly of interest.

Acute Effects

The pathogenic pollutants have been obviously the most important. The effects of pathogenic pollutants are either acute or chronic. The acute effects have been the easiest to determine because of the short time between administration and affliction. The easiest way of measuring acute effect has been by feeding the pollutant in question to test organisms, such as mice or selected fish, and observing the dosage required to kill 50% of the organisms involve. This is usually carried out by feeding a population of test organisms increasing amounts of the pollutant, and for each dosage the number of individuals that succumb is noted. When these data are plotted, an S-shaped curve results. Another way of examining acute effect is to observe the effect of a given pollutant concentration on organisms over a somewhat longer period of time, noting how long it takes for a given dosage to produce mortalities. This may be done by giving a predetermined concentration of the pollutant to the test animals and observing how long it takes to kill half of them. When data from an experiment such as this are plotted, a curve similar to vertical results.

Chronic Effects

There appears to be a dose size for each pathogen below which the pollutant does not have any effect. The concern with threshold values for various pollutants has been one aspect of the study of chronic effects. Chronic effects are the most difficult to measure because of the time involved; in some cases concerns have developed as long as 25 years after initial exposure to the carcinogenic agent. Nevertheless, measurements are made and for many materials a threshold value seems to describe the data reasonably well. Most experiments require some measure of the effect on the organism other than death, so that many different measures are used. These are; deformity in the growth of organisms, damage to particular organs, change in heartbeat or breathing rate, genetic damage to individuals (rather than to the population as a whole), change in the rate of increase of a population, life expectancy of individuals within a particular population, and fecundity of females within the population. Unfortunately, an index may be very descriptive of a pollutant on a particular species, but it may be completely ineffective in describing what happens when this pollutant comes in contact with other species.

Synergism

Another complication in the examination and quantification of pathogenic pollutants has been the phenomenon of synergism. A synergistic effect takes place when the combined effect of two or more materials acting together has been greater than would be expected from the simple sum of the individual effects. Temperature, for example, has been a very strong synergistic parameter, generally enhancing most pollutant effects as temperature is increased.

The same is true for salinity and dissolved oxygen. Some pollutants, on the other hand, tend to decrease the effects of others, as might be in the case of a strong aid and a strong base present at the same time. Other pollutants may make strong pathogens to get precipitated out of the water column or to be collected on sediments, so that in either case they end up on the bottom. Thus it has been extremely important in the analysis of any pollutant to be aware of the other materials present and their synergistic tendencies. Within the marine biosphere may organisms get affected by various pollutants. Whether or not a particular organism has been affected by any given pollutant will depend on many variables. For example, many pollutants occur only in the water column while others, such as some of the pesticides and heavy metals, are adsorbed into suspended sediments and usually find their way to the bottom within a relatively short period of time.

Consequently, these latter types would tend to affect those organisms living on the bottom or feeding from bottom organisms more than they would influence the organisms living within the water column, like free swimming fish. Also of importance in determining the effectiveness of a particular pollutant on living organisms has been its relative solubility in water and oil. Pollutants more soluble in oil will tend to find their way to organisms that have a larger content of oil in their body tissues. This has been particularly true of some of the pesticides which are more soluble in oil than water.

The individual habits of marine organisms also influence their susceptibility to a particular pollutant. Feeding habits with respect to time of day, portion of water column foraged, type of material ingested, and the method of digestion utilized all determine to a great extent the types of pollutants the organism will be exposed to. Similarly spawning habits decide the type of area and time of year in which a species has been particularly sensitive to subtle changes in the environment. These changes influence not only the parents but also the offspring, and may be even their progeny.

Other habits not having to do with spawning also help to determine whether an organism will get exposed to particular pollutants. Whether the organism has been sensitive to light or sound, for example, will often determine whether it gets attracted to a particular outfall. Thus we find that a pure and simple cause effect relationship between a pollutant and an organism is not enough to determine completely the actual response of the organism in the real world.

Once some of the pollutant gets ingested by marine organisms, the major concern of man has been the possibility of this pollutant appearing on his dinner table and causing a pathogenic response. Obviously if a fish ingests a pollutant and a human eats that fish, then that person will get ingested a dose of that pollutant and there will probably be some danger. The question is, "How much?" In order to arrive at an approximation of the actual amount of undesired material finding its way into the reader's stomach, one must be able to follow the pollutant through a number of steps. Most organisms living in the sea tend to concentrate materials existing in the sea to much greater values than they have been in the ocean itself. For example, a diatom, a small form of marine algae having a silicate frustule, is having a body concentration of silicon about 40 thousand times as great as the oceanic waters from which the organism derives all of its material. Thus the diatom concentrates silicon very effectively so that if we are diatoms, a large portion of our diet would be glassy.

Similarly, other organisms, including fish, will concentrate other materials even if these do not naturally occur in the environment. But this amplification or concentration does not stop with one step. We must continually keep in mind the fact that an organism does not exist by itself in the marine environment. Any organism has been dependent upon many other organisms for its existence. An edible fish, for example, might very well subsist on smaller fish which, in turn, might subsist on small zooplankton (small floating forms of animal life), which, in turn, may subsist on phytoplankton (forms of floating plant life). There may be 5 to 10 individual links in a food chain leading to man, and there has been the distinct possibility that for each one of these links a concentration of the pollutant occurs.

Hence, the total pathway to man for the pollutant might be a rather tortuous one, but it might result in a very high conentration of the pollutant material in the fish. This in itself may be of no concern to the individual unless he ingests the fish and if so, how often and how much he consumes. It has been not enough simply to know that there has been some pollutant present in marine organisms; one should also know how much is present, and in addition two other facts have been required. This first has been the accepted maximum level of ingestion of the pollutant before harm results, and the second has been the amount of fish that can be eaten before this level gets reached. In this way it might very well be possible to have a particular kind of fish one or two times a year with perfect safety, while if it has been eaten once a day, it might very well have a toxic effect.

Ecomorphic Pollutants

There are many different types of pollutants which may be classified as ecomorphic, while some pollutants fall into all three of the classes stipulated above and many fall into at least two. Sediment has been a pollutant whose effect has been primarily that of changing the environment since it is not pathogenic and very often is not aesthetically displeasing, although it can be. As it settles to the bottom, it will tend to bury the organisms that live on the bottom, such as oysters and clams, and eventually smother them. Sediment also tends to fill in marsh areas, killing the marsh grass, and completely changing the habitat. As marsh areas are breeding grounds for small fish and an important ecological link in most of the oceanic life, any destructive process such as this has been bound to markedly change the marine biological environment.

On the other hand, in some cases where man's activities are able to produce large amounts of erosion and sediment has been brought down by rivers into estuaries, some estuarine areas have been filled enough to support marsh grasses. In same cases excess erosion can produce marsh lands rather than destroy them. Whenever there has been a depth change in an estuarine or coastal area, the effects have been many. They range from changing the biological environment so that larger fish will no longer live in the shoal areas to affecting the total circulation pattern of the area. If the sediment has been in suspension it also has an effect on the ecosystem, the most obvious aspect being a change in transparency.

By making the water more opaque, light will not penetrate as deeply as before so that the thickness of the layer of water in which plants grow (the euphotic zone) will get decreased. If the water has been fairly shallow to begin with, this decrease in transparency may very well result in the inability of rooted plants to grow and some marsh lands may be destroyed. The effect of sediment, then, has been two-fold; changing both bottom and water properties. Both of these effects tend to change the physical characteristics of the ecosystem. Another ecomorphic pollutant has been oil, although oil sometimes produces toxic effects. The result of oil coating bird feathers has been a well known physical effect. The oil does not poison birds; it simply makes impossible for them to fly, and consequently many of them die. A film of oil on the surface will also decrease the amount of sunlight entering the water and, similarly, will limit the exchange of oxygen from the atmosphere. In addition to the aesthetic and pathogenic effects of oil, they are thus also ecomorphic effects. The toxic effects of various nitrogen ous compounds have been well known but fertilizers and other nutrients washed into the sea possess a nontoxic effect on marine waters. These materials will do just what they have been designed to do: made plants grow, whether these plants be terrestial or aquatic plants.

Consequently, the first effect of a spike of fertilizer added to the sea has been an over stimulation of those plants normally there. Beyond that the undesirable plants (those that are not food for the edible types of marine animals) often get stimulated to grow and these marine weeds grow so rapidly that the overwhelm the other types of plants. As plants have been at the bottom of the food pyramid, the structure and characteristics of the ecosystem have been completely changed, even to the extent of markedly altering the species composition. Once these plants have started to grow at greatly accelerated rates, they may completely cover the surface of the water, disallowing any light from penetrating more than a few centimeters. Those plants on the bottom of the blanket will die, due to lack of light, and those rotting plants will use oxygen in their decay processes.

Even though there have been more plants than before, producing more oxygen in the process of photosynthesis, so many of them are dying and decaying the result has been to decrease the dissolved oxygen. This phenomenon is termed as eutrophication. Solid waste has been another type of pollutant very often disposed of in oceanic areas. In some cases they tend to completely destroy the natural habitat and the result has been not only ecomorphic but also aethetically displeasing.

However, is some cases, the ocean dumping of waste products such as old automobile tires an even automobile bodies has produced an artificial habitat for many organisms that were not endemic to the region previously. However, there are cases where nontoxic solid waste materials have destroyed desirable habitats. Sometimes the dumping of solid wastes makes a change in water flow patterns producing changes in salinity and temperature which influence the biosphere markedly. This usually takes place when the water is reasonably shallow and the volume of solid waste products dumped is relatively large. Another ecomorphic pollutant that seems to be receiving a lot of attention lately is heat which is supplied to the marine environment from electrical power plants or other industrial sources. This heat will increase the water temperature and may result in different kinds of damage to marine organisms.

As most mutations have been caused by thermal activity under normal conditions, the possibility of genetic damage has been real. It is estimated that about 90% of the mutations occurring normally have been thermally caused, while only about 10% have been caused by radiation. Thus with an increases in the temperature, one can expect greater mutation rates or more genetic damage. In addition to genetic damage, heat can cause both acute and chronic problems. Animals can be killed by extreme thermal shock or they can suffer lingering effects and finally expire by being unable to acclimate to the new surroundings. More subtle changes are also noticed as growth rates are changed with increase or decrease in temperature.

Normally a small increase in temperature will make a small increase in growth rate; however, if the temperature change in extreme, this may be reversed. Changes in spawning patterns have also been noted in which fish that normally move upstream to spawn are unable or unwilling to cross a thermal barrier associated with a heated effluents. This may cause the cessation of spawning for a large portion of the species population. Another effect of temperature increase has been to cause certain organisms to grow more rapidly than others, upsetting the previous ecological balance and eventually causing a change in species composition. A similar range of effects to that produced by heat has been caused by salinity changes. These might be produced from the input of rich brine solutions resulting from the effluent of water desalination efforts or a fresh water outfall such as a raw sewage treatment plant. A marked change in salinity will bring about damage similar to that produced by heat, including genetic damage growth rate changes, spawning pattern changes, and changes in species composition. All organisms are used to living within some range of salinity and if this range gets exceeded, they will either die or will move some where else.

Aesthetic Pollutants

The last type of pollutant mentioned above was aesthetic pollutants. Aesthetic pollutants have been those which offend the human senses and as such are regarded by many to be unimportant. In many environmental problem area, though, the aesthetic considerations often receive more attention by environmentally oriented citizens than some of the pathogenic pollutants. This is simply because the aesthetic pollutants have been visible and what can be seen, smelled, or heard appears much more bothersome than that which is hidden, even though the hidden pollutant might be much more dangerous to life.

Consequently, any scheme for controlling marine pollution must include some consideration of the aesthetic pollutants if it is to succeed. An excellent strategy seems to be to focus on the aesthetic pollutants and use them as a mechanism to gather momentum to clean up the rest of the pollution problems. Solid waste material, when piled in shallow areas where it can be seen, has been a typical aesthetic pollutant. Building activities also may be regarded to be sources of aesthetic pollutants. Power transmission cable has caused more than one power company grief because of the environmental consideration given the local citizens who are concerned about the appearance of power cables running across the countryside.

Oil wells, especially the offshore variety visible from resort areas, have been of great concern to most environmentalists. The same thing has been true of equipment required for dredging operations and the spoil areas resulting from these operations. To some people, especially those in living relatively pristine environments near the water, boats can also be an aesthetic pollutant, especially those that cause waves and noise.

Effluents that make the color of the sea to change or increase the amount of suspended sediments also may be regarded aesthetic pollutants. It simply has not been as pleasant to go swimming in water that is cloudy or contains coloring material different form what one would expect in a normal, natural body of water. In essence, any change from the pristine environment appears to many to be aesthetic pollution. Obviously, we are living in an unreal world if we think the marine environment can be maintained in an unspoiled form.

It should be kept in mind that many complaints about aesthetic pollution come from individual who would rather live in the primeval environment not realizing that it would not be available to them unless the commercial world about them supported it. There appears to be a dichotomy wherein we try to maintain the unspoiled environment in certain areas while letting others become overutilized. Nevertheless, there has been a growing movement to include in the cost of every project a relatively small sum to improve its aesthetic qualities. This is, of course, exactly the same philosophy that has been applied to the control of other forms of pollution.

Additional money has been spent in a chemical process, for example, to control the pollution-producing properties of the effluent because society believes that this additional sum of money has been a desirable expenditure. It may similarly be maintained that the expenditure for aesthetic pollution control is desirable as it also will tend to increase the quality of life.

Now that the nature of pollution and some of the sources of this undesirable product of society have been considered, we will now concern ourselves with some of the avenues open to us to accomplish some sort of control. The most obvious control strategy that can be used has been to some-how or other completely prevent the introduction of the pollutant into the environment. This can be carried out by not producing the pollutant or by restructuring the process so that the pollutant is no longer a by product. Let us consider a few major sources of pollutants and some possible methods of control in these particular cases.

Maintening dredging, that is, dredging required to maintain a particular depth, adds sediment to the water column and would not be needed if there were no sediment supply to shoal up major channel areas. This type of dredging has been not a one shot affair, but it has been almost always necessary to periodically redredge the accumulated sediment from these channels and harbour areas. This continual sedimentation has been a normal process which occurs whether man is present or not. In some instances the activities of man tend to accelerate the sedimentation process but this has been not necessarily always true. For example, coastal plain estuaries, often the result of river basin flooding, tend to be a rather ephemeral event in geologic time. They get formed when water level rises, and they disappear with the continual process of deposition of material eroded from the higher reaches of the water shed. Another example of natural erosion has been found on the west coast of the United States in the western most portion of California (that portion where more than 90% of the population lives). This area was formed by erosion of the Rocky Mountains and deposition of the eroded material at the base of the mountains. The most obvious solution to the problem of maintenance dredging has been to cut off the sediment supply to the areas where sedimentation has been a major problem. This can be carried out in a number of different ways. The most effective of all has been to prevent upstream erosion, but at the same time this is perhaps the most difficult.

The entire watershed must be policed very carefully to make sure that the natural landscape gets changed to such an extent that erosion no longer takes place. This implies extensive terracing and replanting of areas that have been denuded for one reason or another. Of course, this also implies that extreme measures must be taken in areas that do not have natural cover, such as agricultural areas or areas where construction is underway. A certain amount of this is being done. Another method of preventing sediment form reaching the undesired are to provide settling ponds. The water is released after it has dropped the major portion of its load and is allowed to proceed downstream. The major disadvantage of this system has been that it is essentially a temporary solution since the ponds will fill up rather rapidly, needing either building of new ponds or dredging of some sort.

Sometimes the sedimentation that takes place in a channel area is a result of sediment loads that are introduced into the marine environment from nearby rather than those carried form far up-stream. This takes place for example, when a channel gets dredged with unstable sides so that the material just tends to slough into the deeper regions. This needs continual dredging as the removed sediment has been almost immediately replaced by bottom material from the sides.

One method of correcting this situations is by stabilization of the bottom. Adding larger, more dense material such as gravel to the bottom has been a technique utilized occasionally, although this has been usually more expensive than dredging so that it has been not done too often. Bottom stabilization, however, is a technique that is being used more and more to prevent the movement of bottom materials within relatively small contiguous areas. Another possibility for preventing the accumulation of sediments in undesirable areas has been to change the natural circulation patterns so that sediments get deposited in an area where they do no damage. This may be carried out in various ways: by changing the physical dimensions of estuarine areas, by putting dams in selected areas, and even by changing the circulation pattern by planting grasses to increase bottom friction.

Another method to be considered in decreasing the amount of suspended sediments has been that involved in controlling the numbers and types or water craft allowed in particular areas. In regions that have been relatively shallow and narrow, high speed vessels tend in churn up the bottom, putting sediments back into the water column that had previously settled out. In addition, large wakes are generally produced that tend to erode the sides of the waterway by the mechanical action of these relatively high energy waves breaking on the shoreline. Both these effects can get decreased by the simple expedient of limiting the speed of boats in certain areas. It has been a fairly effective method; however, if the traffic is extremely heavy it may be necessary to limit the total number of vessels, in addition to their speed. In addition to the general clarity aspect of pollution caused by dredging, there has also been a toxic one. Sediment retrieved from harbour bottoms has been especially likely to contain a relatively high concentration of industrial pollutants such as oil and heavy metals.

During the dredging operation there has been a likelihood of some of this material's finding its way back into the water column, but of more concern has been the longtime leaching from dredge spoil dumping areas. Various methods, such as diking and the construction of artificial islands, are utilized to minimize this leaching but their performance under all conditions has been not too well known. Further research in the handling of dredge spoil gets needed at this time

Another problem that, to a certain extent, appears amenable to prevention at the source is that involving oil. Unfortunately, the sources of oil in the marine environment have been extremely difficult to delineate. All the experts agree that the greatest terrestial source of petroleum products in the ocean is used crankcase oil. Some go so far as to indicate that this source supplies more petroleum to the sea than *any* other source. This may be true because within the last decade or so the habits of American motorists are changed markedly to the extent that a large portion of them now change their own oil, discarding the used oil in a manner that probably makes it to end up in storm sewers.

The major reason for this probably is, the almost complete disappearance of the used lubricating oil re-refining industry in recent years since it was not economically feasible to re-refine oil. However, with the modern continual increase in the price of new oil, the re-refining industry has been having a rebirth so that a large portion of the oil previously thrown away will now be reused. Another entirely different source of oil in the marine environment has been that resulting from accidents that take place during shipping and transferring processes.

Generally speaking, the more handling steps involved, the more accidents and associated losses can get expected. The figures as to how much has been actually lost in handling accidents compared to that lost in normal ship procedures, such that as pumping bilges, has been not well known, but there has been no doubt that a very large portion of the oil reaching the marine environment has been the result of poor handling procedures in both the operation of vessels and the transfer of oil from one carrier to another.

Better procedures are required, but perhaps even more than better procedures, stricter enforcement of accepted safe procedures would go a long way to decreasing the amount of oil that enters the ocean each year. When a tanker empties its load of oil, the dynamic characteristics of this vessel get markedly changed. Its weight gets decreased about a thousand times and therefore it becomes extremely difficult to drive the ship, because it was designed to be driven with a full load.

Therefore, when most tankers do not have a load of oil, they will fill their tanks with water for ballast. When they return for a fresh cargo of oil, they must pump this water from the tanks, releasing large amounts of oil as the residue left in the tanks gets washed out. There are many schemes which are suggested for decreasing the amount of oil that is wasted in the ballasting-unballasting process; however, it is not clear at this time how effective these processes have been or how effectively any requirements can be enforced. Modern supertankers have been extremely large vessels and therefore have major structural weaknesses, so that occasionally they will break up in unusually heavy seas. They have been simply not designed to be utilized under unusual storm conditions, so that there will continue to be losses of this type in rough weather. In passing, it should be mentioned that once oil gets spilled at sea under large wave conditions (usually) the case when accidents occur, there has been no way for this oil to be retrieved.

On the other hand, oil may be retriered in harbour areas where the sea surface is relatively clam, but only when there are waves no higher than approximately 1 metre. The obvious way the only way - to prevent oil from getting to the marine environment when a ship founders has been to prevent the accident from occurring in the first place. Alternatively, once the accident has taken place the oil must be prevented from escaping from the ship. It has been somewhat doubtful whether these solutions will ever be possible.

Most industries are using large amounts of water for cooling purposes or as an integral part of the manufacturing process. Consequently, industrial effluents may be having waste products, heat, leached material from heat exchangers, or even incidental house-cleaning wastes. If the industry, for example, has been using water as a coolant, the heat exchangers must constantly be cleaned to keep the heat exchange efficiency high. These cleaners may be more toxic than process wastes. Consequently, it has been extremely desirable to treat all industrial effluents. Some industries have tried to modify the entire process rather than just clean up the effluent. This often causes an increase in efficiency with a lower over all production cost. If this expenditure can be recouped in a relatively short period of time. This obviously might be expected to be a very cost-effective method of handling unwanted effluents. However, in many cases, there has been no increase in operating efficiency, with the net result being an increase in manufacturing costs passed on to the consumer. Society must then make a decision as to how much it has been willing to pay for a particular product *vis a vis* the amount of pollution introduced in the manufacture of this product.

Sometimes pollution comes from the use of antifouling paint on boats. During the summer time in temperate waters the growth rate of marine organisms get increased markedly, especially those organisms setting on fixed surfaces. An untreated boat hull left in the water for a period of just a few weeks will need enough growth to cut the available speed for a given power input by fifty per cent.

Consequently, all owners of boats left in the water have been very careful to apply some coating to the underside to prevent the accumulation of marine organisms. For large cargo ships or tankers, this could make savings of thousands of dollars in fuel costs. Common practice has been to paint the bottom of the boat with antifouling paint.

Antifouling paint has been so used to leach a toxic material into the water in just the proper amounts so that all organisms setting on the hull will get killed. However, the older types of antifouling paints, those that use copper as a toxin, released enough material to the environment not only to kill possible fouling organisms but also to be harmful to other marine creatures at some distance from the boat. Thus there has been a movement of to improve the quality of antifouling paints by drastically reducing this leaching to toxic material into the water. The solution that has been suggested has been to use new paint vehicles in conjunction with new toxic materials so that the toxic material has been almost completely retained within the vehicle, requiring organisms to come within a very short distance of the boat to be affected. One result of recent research is to develop a family of paints using acrylic resins as vehicles and tri-butal tin flouride (TBTF) as the primary antifouling ingredient. In the first few years of use this paint holds great promise because it is less toxic in the water, lasts longer, and does not allow bottom growth more effectively. In their very effectiveness these new paints pose a problem, because the toxin gets retained within the paint for a longer period of time.

Paint scrapings have been now more toxic than they were with the old copper bottom paint. This means that it has been somewhat more dangerous to strip the paint from the bottom of the boat when applying a new coat and precautions must be used to prevent the old paint from getting back into the water environment. When sandblasting large ships, for example, precautions have to be taken to retain the sandblasted material rather than allowing it to be simply washed away. Thus the solution of one problem presents another, but these problems have been solvable if some care has been taken to see that additional health hazards do not get created.

Some of the juicier pollution scandals that have taken place in recent years have involved various pesticides and the problem has been that many of the chemical compounds get introduced for public sale before the total effects of these materials have been known. It seems to be almost an insurmountable problem to determine what the total effects of a new compound will be before it has been put on the market because so many new products are introduced every year.

However, a large portion of the problem has been undoubtedly due to basic carelessness. In both the manufacturing and use of these materials proper precautions are not being taken into consideration. Many users, for example, undoubtedly believe that if a certain amount per unit area will suffice, twice as much will produce twice as much control of either weeds or bugs. This excess toxin is washed into the streams and ends up killing marine organisms rather than protecting plants. It would be evident that more effective control of both the manufacture and use of pesticides and herbicides would go a long way toward eliminating undesirable side effects.

The disposal of solid waste has been very necessary and there even seems to be a fear that if the solid water problem is not solved, eventually there will be no room left for people. Because of this fear, large amounts of solid waste get dumped in the ocean. Some of these solid wastes take the from of sludge from sewage treatment plants whereas others consist of large pieces of discarded consumer products. In the latter case the problem might very well solve itself as the cost of scrap metal continues to increase. One possible use for sewage treatment sludge has been in refurnishing land disfigured by strip mining. Often the overburden removed to get at the coal seam has been not capable of supporting life nor it has been even particularly stable.

The addition of sludge from sewage treatment plants is considered as a method of soil treatment to make it possible to support a viable plant population. It is a concept that has not been tired yet but one which will be tried when it becomes economically feasible to transport the sludge to the strip mine. Another possibility for using the sewage sludge, if it contains only nutrient materials, has been suggested and has been considered by some of the Middle Eastern oil producing countries. The idea has been that empty tankers be filled with sewage sludge for the return voyage rather than with ballast water. This sludge will then be utilized to bring the desert regions back to a point where they have been capable of supporting a large agricultural industry. Other forms of solid waste being dumped in near shore areas have been dumped there primarily because there has been no other acceptable depository.

One method of decreasing the severity of this problem has been simply to decrease the amount of solid waste produced by society. Packaging of consumer items, for example, seems to be a good place to start. From a disposal point of view the catsup bottle is probably the worst possible package. The package weighs more and takes up more space than the contents. The Navy has done excellent work in decreasing package size and weight because there are definite limitations aboard ship, but not too much has been done with the private sector of society simply because there hasn't been enough action on the part of consumers to insist upon small volume and weight packages.

One possible exception to this has been the introduction of lightweight plastic bottles for large size drinks to replace the heavyweight glass variety. They are somewhat thinner than glass, weigh less, and may be incinerated, so the disposal problem has been somewhat alleviated. However, with the continuous increase in petroleum costs, plastics might not remain economically feasible in the future.

It has been found that biological communities living in the sea need a certain amount of physical shelter. Some species just like to be alone, whereas in other cases the young of the species require places to hide to protect against predation. A marsh or wetlands area has been a good breeding place because both of these conditions prevail. It is found that in many cases these conditions can be created artificially by introducing large junk articles such as old automobiles or worn out ships to form effective artificial fishing reefs. There has been disadvantages to using waste products for this sort of activity. One has been that in many cases these junk, objects get transported to undesirable areas by dynamic oceanic forces.

Pollution has been not controlled when automobile fenders and doors wash up on swimming beaches. When creating a finishing reef, steps must be taken to retain the integrity of these reefs for a long period of time. If proper attention has been paid to the magnitude and nature of oceanic forces, artificial reefs night be made relatively permanent. With the present state of the economy where things are some what more expensive, it appears that more items will find themselves in the near future on the list, of materials to be recycled or reclaimed.

Industrial processes will alter as it becomes more feasible to extract materials from the waste stream from a cost standpoint and, at the same time industrial. Waste streams will become less and less polluted. Even in the area of multiple waste product use, like the combination of waste heat from either manufacturing or electrical power generation and waste nutrients from sewage treatment plants for the purposes of aquaculture, cost will always be the controlling factor. It is known that elevated temperatures and the artificial addition of nutrients will increase growth rates of many organisms markedly.

The choice of which organism to grow has been involved not only with which would be best to grow, but also which there has been a market for. Setting up an aquaculture system involves picking organisms which respond favourably to the environment, developing a system by which the optimum growth rate can get produced, developing a marketing system so that the end product will be sold at the maximum possible price, and assuring the continuing existence of a market for everything produced. Not only has been science included but also sociology and economics must be considered, and it is probably in these latter two areas that the major difficulties reside. Hence, no sophisticated aquaculture systems employing both waste heat and nutrients have been developed at this point, but it has been probably only a matter of time before we will have extensive aquaculture using man's waste products.

Another method of contorlling pollution has been to store them at some distance from man's activities so that they have been as inaccessible as possible to the major fraction of the population and pathways to man. Occasionally nuclear waste products are being dumped in deep oceanic areas in the hope that when they do break out of their canisters, they will not find an easy pathway back to man because the deeper life forms have been probably not in any food chain involving man. Perhaps not so well known has been the fact that deep ocean disposal is used for other materials such as average ordinance and poison gases for which the military has no further use. In the case of nuclear products, there has been a very vociferous school of thought that believes storage facilities should allow waste retrieval if a technological break through is ever accomplished that would allow these nuclear waste products to be profitably utilized.

Asimov's suggestion has been to implant our hazarodus waste products in the oceanic trench areas so that within a few thousand years they will get carried into the interior of the earth and no longer be bothersome. The only problem with this suggestion has been that the exposure time before the canister having the waste products is effectively buried in the crust is relatively long.

During this time the canister seal has to be absolute Whether this can be accomplished with present technology of presently being debated within the scientific community. Nevertheless, storage has been still a viable method of controlling pollution, especially for the more toxic materials.

Another method of controlling pollution has been by specifying certain areas that can be utilized for effluent discharge and other areas that cannot. Although we are even starting to convert our city dumps into usable areas, such as golf courses or ski slopes by covering, planting, and land-scaping as rapidly as possible, but it will be some time before ocean dumping areas are reclaimed in this fashion. Certain coastal regions would be reserval for industry, others for sewage treatment, while still others would be set aside for recreation. Some of these activities have been such that they may get pursued in conjunction with one another, whereas others cannot.

The idea has been for every coastal area to try to get the optimum utilization of marine resources available. In order to accomplish this type of zoining most effectively, a great deal more knowledge is needed than is presently available about the synergistic effects of various types of activities. One of the major advantages of optimum zoining has been that it would allow maximum avoidance of undesirable synergistic effects. The biggest stumbling block to coastal zoining probably has been inovercoming enternched interests in regions where the optimum activities have been found to be different from those presently underway: a socio-economic problem rather than a technological one.

The concept is some what interesting one, because it essentially permits some pollution of certain areas while refusing to permit any pollution of other locals. This has been different from the way pollution has been presently controlled, where the effort has been made to limit pollution in all areas to the minimum possible amount. With the joining concept pollution would still be limited as much as possible, but it might permissible to allow the water quality of ocean to degrade below the mean in some places, because in others it would be higher than the mean.

One last method of pollution control is taxation which can be utilized to encourage or discourage certain activities in addition to raising funds to support government programs. Pollution would be allowed, but it would be taxed at a rate proportional to the environmental insult. Some economists suggest that a tax on pollution will simply be a license to pollute and will therefore encourage pollution rather than discourage it. From a management point of view there have been a number of different directions that can be taken to control pollution, including both technological and socio-economic alternatives. So far society has tended to aim more in the direction of the technological alternatives. But there appears to be a definite swing toward other methods. In the years to come it might be possible that more and more emphasis will be kept on these non-technical methods, especially as the tradcoffs, costs, and benefits become better understood.

It has been said that pollution is a biological problem that can be solved by the application of physical and chemical methods. Whether this is a completely true statement is often a moot point, however, because the opportunity to apply any solution is usually won only after the employment of politics. Thus the study of pollution is a truly interdisciplinary one, including as it does the three Ps-pollutants, processes, and people.

Effective control of marine pollution requires further research work aimed at evaluating the extent of pollution of marine ecosystems, determination of the biological effect of toxicants and prediction of biological consequences. However, such an approach alone is inadequate, since the problem is urgent and suitable measures must be taken as soon as feasible. Small marine organisms have a low resistance to toxicants. This fact was utilised to develop a system of rapid bioassays for immediate evaluation of the harmful effects produced by any given substance present in sea water. The more promising applications of the results and methods toxicological research to the protection of the marine environment from pollution include: standardisation of the components of various toxicants in the sea based on biological criteria; ecotoxicological characterisation of sewage effluent and its components, for the determination of methods and norms of permissible dumping; recommendations on the priority and stepwise sequence of reducing the volume of sewage discharged by various industries.

Oil Pollution

Pollutants originating from domestic and industrial sources and from agricultural operations due to high inputs of heavy metals, hydrocarbons (crude oil), pesticides fungicides, insecticides have spread to coastal areas through rivers run off, via the atmosphere, direct discharges or dumping into the sea where the most of the active primary and secondary production direct, are being polluted, causing serious damage to acquatic ecosystems.

Oil Spills

Oil spill on ocean waters kills birds, fouls beaches, and poisons marine life. Hasan Jawaid Khan gives an account of these deadly oil spills, the damge they bring about and the ways by which an oil slick is contained tense of thousand of sea birds acted with oil shivering and dying on the beaches. At least a thousand sea otters succumbing in the oil with their livers and kidneys fatally impaired. And an unknown number of seals, whales and porpoises poisoned. Even before the tragic images of the deadly oil spill on the Alaskan coast could fade from memory, the specture of an oil spill is back again to haunt us-only this time it is much too bigger. The enormous oil slick in the Gulf region, 30 times more than the Alaskan episode, threatens to set into motion an unprecedented ecological disaster that may take decades to fix. Ironically, the oil that drives millions of vehicles round the world, sometimes drives countless marine animals to a most cruel death, crude oil when refined drives automobiles and airplanes, runs factories and farm equipments, provides gas for heating and cooking and is a source for making drugs, cosmetics and fertilizers. But when crude oil spills on to ocean waters it spells death for marine animals and disaster for the ocean ecology. There has always been a danger of major oil spills accuring near a coastal area with supertankers being increasingly used to carry industrial and heating oils. Several of these huge ships have been grounded and cracked open by the seas, their vast cargoes spilling out heavy oil into the oceans where currents can carry it many miles before it disperses and gradually sinks to the floor. Sometimes the currents carry the oil toward land where it fouls beaches and kills marine life. The world first woke up to the disastrous consequences of an oil spill when on 18 March 1967 a Librarian tanker. Terry Canyon, ran a ground on the southwest cast of Great Britain, near the entrance to the English Channel, spilling 60,000 tons crude oil into the sea. Oil splittered on to 160 kilometers of coastline killing countless number of fish and birds. Two years later, in January 1969 occurred the second major oil spill off the coast of Santa Barbara in the United States when an offshore oil well below out resulting in the discharge of oil at the rate of 1,000 gallons per hour causing extensive damage to the coast. In 1978, in the much publicized Amoco Cadiz disaster 68 million gallons of oil was dumped all along the French coast. And then on March 24, 1989 the supertanker Exxon Valdez belonging to the Exxon corporation, the largest petroleum company in the world, plied on to a reef off the coasts of Alaska. Through five gashed in its hull the 30,000 ton super tanker spilled over 11 million gallons of oil into the clean water of Alaska's Prince William Sound. As the oil hit 1,930 kilometers of shoreline. 100,000 seabirds died including 150 rare species of baldeageles. An unknown number of dead seals sank to the bottom of the ocean. And at least 1,000 sea otters perished. Even a few deer and a couple of bears that fed near the shoreline were found dead. Closer home, Indian coasts have also been witness to a few tanker disasters, none too serious thought. In July 1973, 3,000 tons of oil washed on to the Gujarat coast when an oil tanker, cosmos pioneer, ran around. In 1974 an American oil tanker Transhuron colided with one of the atolls of the Lacadives spilling 5,000 tons of special furnance oil causing extensive damage to marine life. Similarly destruction was witnessed when the oil tanker Lajpat spilled thousands of tons of oil along the Bombay coast. The last time an oil spill threatened the India coast was in June 1989 when a Malteese tanker M.T., Puppy collided with a British vessel. The Maltese tanker spilled over 5,500 tons of furnance oil into the open seas off Bombay. At that time the fear of contamination of fish brought the fish industry to virtual halt at people stopped buying fish for fear it might cause cancer. But not all the petroleum that gets into the marine environment is through oil spills. Natural seeps have been discharging petroleum into the marine environment for millions of years in amounts substantially greater than those resulting from present offshore production activities. More than 200 submarine oil seeps have been identified worldwide. An assessment of petroleum pollution of the world's oceans by the national Academy of Sciences, Washington, estimated that between 1.7 and 8.8 million metric tons of oil enter the oceans annually. Oil spills are by far the major contributors to this pollution. While other categories of pollution have shown a fifty per cent decrease, oil spills have been on the rise. But what is it that makes crude oil shore so deadly? Crude oil is rarely used in the form of product at the offishore wells but it is converted in refineries into a wide range of products such as gasoline, kerosene, diesel fuel, jet fuel, domestic and industrial fuel oils and petrochemicals feedstocks. Until the oil is refined it contains potentially dangerous components.

Crude oil is primarily made up of compounds of carbon and hydrogen called hydrocarbons. These hydrocarbons may be paraffins parafin oil is used as a fuel in heaters and lamps cycloparaffins (naphthenes) and aromatic compounds in varying proportions. Crudes founds in easter and aid western sections of the United States are predominantly paraffinic, while those found along the Gulf coast are usually naphthnic. Crude oil also contains varying amounts of sulphur compounds, a small amount of nitrogen and very little oxygen or olefins which are unsaturated hydrocarbons. Nickel and vanadium are present in all crudes in high concentration. Iron may be found in inorganic form as a result of corrosion of pipes and containers used to produce and transport oil. Paraffins like methane and ethane are asphyxiants, that is, cause suffocation. Some paraffins are central nervous system depressants. Liquid paraffins can remove oil from exposed skin and cause dermatitis or pneumonia in lung tissue. However, these are the least toxic of the hydrocarbons. The effects of cycloparaffins are much like those of the paraffins except that unsaturated cycloparaffins are more toxic than the saturated ones, breathing high concentrations of their vapour can result in Irritation and an aesthesia, the most toxic of the hydrocarbons are the aromatic hydrocarbons. Inhalation of the vapour can cause acute intoxication. Benzone is particularly toxic and long-term exposure can cause anemia and leukopenia even at low concentrations. Sulphur compounds present in crude oil can also be toxic. Carbonyl sulfide is dangerously poisonous. It is lethal to rats at 2900 ppm (parts per million). The mechanisms of toxic action appears to involve its breakdown to hydrogen sulfide. It acts principally on the central nervous system, with death resulting mainly from respiratory paralysis. Sulphur in the form of aromatic thiophenes, benzothiophenes, can be moderately toxic. These damage the livers and kidneys. Certain sulphur compounds like sercaptan can also be lethal. Although the long-term effects of an oil spill are not so well known, the Alaskan oil spill provided an opportunity to study the short-term effects. As the 11 million gallons of oil spilled by the Exxon Valdex hit the shores, seabirds started shivering and dying. The oil had broken down the natural insulating oils and waxes which shield the birds from water. As a result hypothermia (abnormally low body temeprature) set in and the birds started dying. About 150 rare species of bald eagles also became victims when they ingested oil, seaven giving on oily seabird carcasses. Cleaning away the oil from the birds has not been very successful as it causes a severe strain on the birds. However, the British Wildlife Rescue Association has designed wool sweaters which absorb oil from the stricken birds. The sweaters have already been used to clean birds in the North sea and were found to be more effectively than cleaning with detergents. About 1,000 sea others died after their far became saturated with oil and provided no insulation. Others developed respiratory ailments as volatile components of the oil weakened membranes in their lungs. Some suffered liver and kidney damage caused by ingesting oil while cleaning their coasts. Firstly were found to be the least affected. When fish from the oil spillage sites were tested no hydrocarbons were found in their flesh. The fish are efficient at converting hydrocarbons they ingest into metabolities which are then everted from the liver to the gall bladder. Moreover, because of their mobility the fish escape large-scale contaminated as they are able to avoid heavily contaminated areas. When oil spills on to ocean waters it spreads rapidly over the surface. The force of gravity causes the lighter oil to seek a constant level by spreading horizontally on the heavier water· The movement of the oil is caused by the combined action of wind, surface currents waves and tides. Oil on water normally moves at about 1/30th of the wind speed. With large amounts of oil, as with accidental spills, the oil drift is largely independent of spill volume, sprading, or weathering. Since the thick portion of the slick drifts faster than the thin portion a heavy oil accumulation forms the leading edge of an advancing slick.

As the oil spreads it comes under the influence of several weathering processes that ultimately break down a spill with time. The first of these is evaporation which is dominant soon after oil is released. Evaporation is determined by wind velocity, nature of the roughness, and temperature. Almost 50 per cent of oil is lost through evaporation during an oil silk's lifetime. Some of the light, low-bowling hydrocarbons, such as benzene, toluene, and xylene are lost through evaporation. Their removal decreases toxicity of the oil to marine life. Muce of the oil that evaporates is photooxized in the atmosphere and some of it may return to the seas as atmospheric fallout. A very small fraction of the oil dissolves in the water in the early stages of an oil spill. Lower-molecular-weight compounds tend to be the most soluble. However, this process is very insignificant. Natural dispersion also removes some oil from the water.

In this process, small droplets of oil which are larger than dissolved molecules, are incorporated into the water in the form of dilute oil-in-water suspension. Crude oil contains trace amounts of altrogen, sulfur, and oxygen-bearing compounds which can act as natural surfactants. These surfactants reduce the oil-water interfacial tension allowing the oil to break up and to disperse into droplets more readily. A part of the oil is also broken down by microbial degradation. Microorganisms present in the water have a great capacity to utilize hydrocarbons present in the oil as in energy source. This process especially speeds up when the oil disperses. Some organisms may also ingest dispersed oil droplets and subsequently deposite them as fecal pollets. Emulsification is yet another weathering process in which water is incorporated into the floating oil forming a water-in-oil emulsion. Such emulsions, which may contain from 20 to 80 per cent water, are often very viscous and referred to as "mousse" once mousse formation starts, cleanup operation run into difficulties as it is almost impossible to pump.; Ultimately the mouse is broken up by wind and wave action into tar balls that may wash ashore.

However, breaking up of an oil spill takes time. And the longer the oil continues to flat on water the more damage it docs to the marine environment therefore, cleaning to an oil spill fast assumes prime importance. An oil spill can be cleaned up by mechanical removal of the oil, by the use of chemical and microbial surfactants, by using absorbants, or by burning the oil slick. Spills can be dealt with more easily if they are confined to a small area on the water surface. For this, mechanical booms or barriers are spread around on oil slick to check its progress and prevent it form hitting the shoreline. But sometimes high waves may splash the oil over the booms. The oil may even pass below the booms once it has piled against the barrier. So after the oil has been contained by the booms it is sucked up from the surface of the sea by vessel known as skimmers. Skimmers now available generally fall into one of two categories. The first, mechanical surface skimmers, remove the top layers of the water and oil from the surface but they are not very effective if there are high waves. They gulp large amounts of water unless some provision is made to allow the weir suction port to follow the water surface. A second type of skimmer operates on the principle of selective wetting of a surface by oil rather than by water. Rotating metal disk or conveyor belts dip into the water, a surface layer of oil removed. The spilled oil can also be treated with dispersants. These chemicals are sprayed from aircrafts or ships. Dispersants cause the oil to spread further and disperse in a manner similar to the way soap removes oil from one's hands allowing the oil to be emulsified and washed away with water. A dispersant contains a surfactant, a solvent, and a stabilizer. The solvent usually comprises the bulk of the sispersant and enables the surface-active agent or surfactant to mix with, and penetrate into the oil slick and form an emulsion. The stabilizer fixes the emulsion and prevents it from coalescing once it is formed. Dispersion increase the slick surface area and allows a rapid increase in the rate or microbial decomposition. But although dispersants are effective in diffusing a slick, these chemicals are toxic to marine life. And, since the emulsified out droplets sink into the water body, oil stays in the waters with its toxic components such as benzene and toluene entering the food chain with disastrous long-term effects. Absorbents are also used to facilitate the cleaning up of oil spilts. When they are applied to the slick, they absorb the oil and prevent it from spreading, and when the absorbent material is removed from the water, the oil is removed.

Natural materials such as peat moss, straw, sawdust, and pine bark can be used. Synthetic absorbents include polyethylene, polystyrene, polypropylene, and polyurethane. Of all synthetic absorbents, polyurethane is the most promissing. Burning oil slicks on the open seas has generally been less successful because the more volatile light fractions evaporate quickly from the oil slick. Also, the water removes heat faster than it can be created to support the combustion. But burning oil slicks also leads to extensive air pollution. By far the safest way of rating an oil slick is by using biological agents for degrading oil. These microbial surfactants are sprayed from the air. They mix with the oil, emulsify it and disperse it throughout the water body so thinly that it no longer remains hazardous. However, not every clean-up method can clear up all oil spills. Clean-up methods are chosen on a case by methods are chosen on a case-by-case basis. But whatever the method, it is important that the efforts are mounted as soons as the oil spills is discovered. For the longer it takes for the cleaning up to start, the more distance the oil travels. And as the oil charts its destructive course it fouls beaches, kills marine animals and contaminates the food chain.

Gulf Oil Slick Making History

According to H.J.K, the massive oil slick in the Gulf is making history. With the total spillage of oil in the Gulf waters now standing at more than 330 million gallons, it is already 30 times more than the quantity spilled by the Exxon Valdez on the shores of Alaska. The oil slick that now stretches over an area more than 80 km long and 20 km wide is moving south at a speed of 20 km a day. Having already hit the Saudi shores, affecting birds and marine life, the slick is now threatening the shores of Bahrain, Gatar and the United Arab Emirates. Apart from killing marine life the oil also threatens to disrupt the deslination plants on the western coast of the Gulf that provide nearly half the water to Arabian states. The enormous oil slick will disrupt the already fragile ecosystem of the region. According to the world conservation Monitoring Centre in Cambridge, between 1 million and 2 million birds will die as a result of the gulf oil spillage. Some species may never recover. Cormorants coates with oil have already started dying on the Saudi beaches. Dolphins may escape contamination as they seem to be able to sense a spill and street clear of it. The dugong or sea-cows will be affected with their food supply of sea or grass getting contaminated. The three already endangered species of turtles the green, leatherback and hawks bill may be seriously affected. Coral reefs housing a diverse variety of fish maybe destroyed. The oil is likely to affect the mangroves of Iran and a number of valuabale shrimp fisheries. No immediate threat to the Indian coastline is perceived so far according to oil experts. The oil split from the Al-Ahmadi terminal is export quality Kuwaiti crude, extremely light and fluid. Hence, much of it would evaporate. In the first 24 hours, 35 per cent of the crude is likely to evaporate and in the next 10 days about 45 per cent. Even if some of the oil manages to find its way out of the Gulf, much of it would be broken up by the rough tropical waters of the Arabian Sea. Due to burning of the oil in the Gulf and attempts for its removal, a very small quantity of tar balls may eventually hit the Indian coast. These tar balls can just be collected and set on fire.

Biman Basu made survey on Gulf oil spill and he pointed out that the oil spill in the Gulf is the largest so far, spreading over 700 sq km, it does not pose any pollution threat to Indian waters. According to Dr. Peter. "This is not so much environmental terrorism as an environmental crime." said Worldwatch founder Less Brown, "We need a code of ethics like the Geneva Convention (governing treatment of prisoner of war) for the environmental filed. The Gulf could die in a matter of months." Ancient warriors knew how to use the environment as a weapon. Corpses were catapulated into besieged cities to spread disease; water supplies were diverted to cause drought or floods; wells were poisoned. The Romans spread salt on the fields of Carthage to deny food to their enemies; Russian troops defeated Napoleon's army by torching the earth as they retreated, destroying all crops animals and human dwellings. Prime Minister Winston Churchill planned to spread burning oil on Britain's beaches to block any German invasion in the Second World War. In the same war the Dutch opened up their sea dikes to flood their own land in a bid to stop the advancing Nazi troops. The U.S. Forces in Vietnam also practised environmental warfare when they sprayed Agent Orange to defoliate vaste stretches of forest land deprive Viet Cong guerillas of protective cover. The spraying destroyed large number of animals that lived in the forests, polluted waterways with the reputedly, cancer-causing chemical dioxin, and Vietnamese exposed to defoliated areas reported increased rates of miscarriages and birth deformities. Heydemann, Science Counsellor in the US Embassy in New Delhi, the oil may not come out of the Gulf for years because of the current patterns in the region. By that time most of it would have been disintegrated by natural bacteria. But the slick will cause irreparable damage to the rich marine ecosystem of the Gulf.

Ben Barber studied that the gaint oil slick released into the waters of the Gulf during the war has been denounced as "environmental terrorism" which along with deliberately set oil well and refinery fires, could alter the region's ecology for the years to come. Yet, since Biblical times, armies have destroyed the environment to deny and enemy food, water or shelter. "The phenomenon is not new there have been efforts to degrade the environment as part of war fare for sometime," said Mr. Chris Flavin an analyst with the Washington based environmental think-tank, Worldwatch institute. But modern technology has upped the ante", he added "There's a much greater impact." The estimated 460 million gallons of oil released in to the shallow waters of the Gulf is by for the world's greatest oil spill. Spreading south from off the Kuwaiti and Saudi coasts towards Bahrain and Gatar, it threatens millions of sea birds and marine life forms, including the rare dugong or sea cow, the mammal that supposedly gave rise to the legend of the mermaid. Since the war ended in 1975, low-value bamboo and grasses have invaded defoliated areas, preventing the regrowth of valuable hardwood forests. The U.S. also seeded clouds over Asia to try to alter Vietnam's rainfall and make supply trains to muddy to use, said Prof. George Rathjens of the Masschusetts Institute of Technology. A political science teacher who has specialised in nuclear weapons and environmental issues, he said some modern military planners have proposed cloud seeding to divert rainfall from an enemy's crops. In addition to the releases of oil into the Gulf, several Kuwaiti oil refinerises were set on fire in what may have been an attempt to raise a smoke shield against attacking aircraft. Some scientists have warned that such huge fires could block out sufficient sunlight to disrupt monsoon rains as far away as South Asia. The worst-case scenario warned that massive famine could result. Prof. Rathjens, however said it is unlikely the fires would result in the phenomenon known as "Nuclear Winter" the predicated effect on the planet of an atomic war.

Ballerini-D and J.P. Vandecasteele studied the aims essentially at quantifying the processes of microbial degradation of crude oil in optimal conditions compatible with a marine environment. Our purpose, specifically, was to determine the proportion of crude which could undergo biological degradation, and to improve our knowledge of the biodegradation kinetics of the various hydrocarbon families which constitute a crude oil. This work was performed in laboratory reactors where various physiochemical parameters such as pH temperature, stirring speed and air flow rate were strictly controlled. The study was conducted first in batch then in continuous cultures. The composition of the mineral medium used for the culture is shown in Table 1. This composition was chosen by taking the average composition of the Atlantic Ocean as a basis (1) and enriching it in nitrogen, phosphorus and iron. The concentrations of these three elements were respectively raised from 0.5 to 235 mg/L for nitrogen, from 0.007 to 26.7 mg/L for phosphorus and from 0/001 to 0.4 mg/L for iron and the pH of this medium was adjusted to 8.1 the average pH of seawater.

Table 1

Mineral Medium

Salts	Concentration (g/L)
Sodium Chloride	23.9
Megnisium Chloride	10.8
Calcium Chloride	1.15
Potassium Chloride	0.74
Sodium Sulphate	0.40
Sodium Sulphate	0.023
Sodium Bromide	0.117
Sodium Fluoride	0.002
Sodium Bicarbonate	0.196
Silican Dioxide	0.022
Sodium Oxide	0.022
Ferrous Sulphate	0.002
Amonium Chloride	0.9
Potassium Hydrogen Phosphate	0.15

In order to approximate the conditions existing at sea, where evaporation is not negligible (2) we mainly used crude oil which had been topped at 240 degree C. In the present case, we used a residue of an Arabian light crude, obtained by distillation under reduced pressure, in which all the fractions distilling below 240 degree C had been removed.

Table 2

N-Alkanes Composition of Saturated Hydrocarbons

C-Number	Arabian Light 240+	C-Number	Arabin Light 240+
15	3.44	26	3.75
16	7.72	27	3.18
17	1013	28	2.54
18	10.42	29	2.05
19	9.73	30	1.85
20	9.10	31	1.47
21	8.23	32	1.20
22	7.03	33	1.01
23	6.12	34	0.76
24	5.37	35	0.52

Table 2 shows that in the alkane distrubution of Arabian light crude oil topped at 240 degree C n-paraffins of chain length below C 14 have been completely removed. In oil spills at sea, all compounds with a number of carbons below 15 disappear after a residence time of about 10 days. The reactor used in our studies is glass cylinder equipped with two stainless steel end plates. Its total volume is approximately 3 liters. Stirring is achieved with a simple 4-bladed turbine, the speed of which is adjusted between 600 and 800 pm. The reactor is equipped with automatic pH of the culture is maintained at 8.1 by addition of 1N sodium hydroxide, the temperature is kept at 20 degree C = 02 degree C and the air flow rate is maintained at a value of 1 liter air/L of medium/hr. Experiments are conducted in nonsterile conditions. The analytical methodology utilized to follow the processes of crude oil degradation is also summarized. Gaseous effluent from the fermentor first goes through a flask containing carbon tetrachloride where the hydro carbon vapours are trapped, then in a second flask containing a known amount of IN potassium hydroxide, which traps carbon dioxide. Hydrocarbons are then determined by infrared spectrophotometry and the carbon dioxide is determined by titration. Separation of asphalotenes is first performed on the hydrocarbon residue. Asphaltenes are preciptiated by boiling with heptane under reflex for 1 hour. Insoluble materials are then separated by filtration, rinsed with hot heptane, then dried and weighted.

On the residue obtain evaporation, separation of the three main classes of compounds of crude oil, saturated hydrocarbons, aromatic hydrocarbons and resins is performed either by thin layer chromatography on silica gel or by liquid chromatography on a silica gel column using cyclohexane as the elution solvent. The sum of the weights of the four fractions recovered in this procedure always represents 90 to 100% of the intital weight of the total oil sample. On the saturated and aromatic fractions, more detailed analyses using gas chromatography and mass spectrometry are then performed. The isolation and selection of the strains of microorganisms used were performed starting from marine silt and sand samples collected at places hit by crude oil spills. The first enrichment cultures were performed by successive transfers in liquid medium to which 2 g/L of crude had been added, first in shaken flasks and, soon afterward, in laboratory fermentors in order to obtain perfectly defined culture conditions. In the course of these successive batch cultures we observed a very important decrease of the lag phase and an improvement of the final biodegradation performance. A large improvement of the dispersion of the hydrocarbon phase in the mineral medium was simultaneously observed. A very fine emulsion of the organic phase in the aqueous phase was thus obtained, leading to a marked increase of the reactional area. The bacterial association present in our selected cultures consists of four dominant bacterial genera: *Pseudomonas, Moraxella,* A *Cinobacter* and *Flavobacterium.* During a batch culture we studied the changes in the bacterial flora. Although at the beginning the dominant strains belonged mainly to the genus *Moraxella,* the *Pseudomonas* strains predominated later on, as oil biodegradation advanced. The biodegradation process during the batch cultures can be illustrated by following the changes of various parameters. For example, we have represented biomass and carbon dioxide production, the latter in both the aqueous and gaseous phases. Total production of carbon dioxide follows an evolutaion quite comparable to that of biomass production. At the end of the culture, whereas hydrocarbon consumption reaches 2 g/L the total amount of carbon dioxide produced is 1.45 g/L and that of biomass is 1.4 g/L. The changes in time of dissolved oxygen concentration in the medium, as well as consumption of ammonia nitrogen and hydrocarbons are shown in figure 6. The amount of hydrocarbons lost by evaporation is quite negligible as it is measure on a 48 hour period and is always below 10 mg. A perfect correspondence between the evolution of the consumption of ammonia nitrogen and hydrocarbon can be again noted. In addition, during the acceleration period of the biodegradation process, a very distinct decrease in the oxygen dissolved in the medium occurs. Actually, dissolved oxygen concentration even becomes a limiting factor since values of 5% of the saturation level are read. Later, as the velocity of the bio-degradation process decreases, the concentration of oxygen dissolved in the medium increases again, but without reaching the saturation value.

Table 3
Degradation of Saturated Hydrocarbons

Hydrocarbons	Start of batch [mg/L]	End of the batch [mg/L]	Hydrocarbons consumed [mg/L]	%
Saturated Hydrocarbons	1170	390	780	67
Alkanes	739	82	657	88.9
1—ring	99	51	44	44.4
2—ring	136	89	47	34.5
3—ring	92	79	13	14.1
4—ring	55	50	(5)	—
5—ring	27	24	(3)	—
6—ring	22	15	(7)	—

Table 4
Degradation of Aromatic Hydrocarbons

Hydrocarbons	Start of batch [mg/L]	End of the batch [mg/L]	Hydrocrbons consumed [mg/L]	%
Monoaromatics	268	138	130	48.5
Diaromatics	251	168	83	33.0
Other compounds	481	424	57	11.8

Table 5
Degradation of Monoaromations

Hydrocarbons	Start of batch [mg/L]	End of the batch [mg/L]	Hydrocrbons consumed [mg/L]	%
Alkylbenzenes	113.4	36.6	76.8	67.7
Benzocyclo Paraffins	100.0	53.8	46.2	46.2
Benzodicyclo Paraffins	54.9	47.4	7.5	13.7

Table 6
Degradation of Diaromatic

Hydrocarbons	Start of batch [mg/L]	End of the batch [mg/L]	Hydrocrbons consumed [mg/L]	%
Naphthalenes	98.3	49.1	49.2	50.0
Acenapthenes	81.4	60.4	21.0	25.8
Fluorenes	71.0	58.3	12.7	17.9

The results of mass spectrometric analysis of the saturated fraction are presented in Table 5. First they indicate that alkanes are the main target of biodegradation since 88.9% have disappeared at the end of the culture. This allows us to conclude that, in addition to *n*-alkanes and isoprenoids which represent only 14.8% of the saturated fraction, the great majority of isoalkanes is consumed by microorganisms. Among naphthenic compounds, 1-ring and 2-ring naphthenes are the main targets of biodegradation, with respective percentages of biodegradation reaching 44 and 47%. For the aromatic fraction results presented in

Table 5 indicate that the microbial action is particularly evident at the level of mono—and diaromatic compounds. At the end of the culture all mono and diaromatic compounds with a carbon number below 16 have disappeared.

Among monoaromatic compounds, it can be seen in Table 6 that the compounds most susceptible to microbial action are alkylbenzenes, 67.7% of which have disappeared at the end of the culture, and benzocycloparaffins, 46.2% of which have been consumed during this experiment. The results presented in Table 7 illustrate the microbial action on diaromatic compounds and, in particular, the effect of microbial attack on the residual content in naphthalene derivatives, 50% of which are consumed during the biodegradation process.

The primary purpose of this paragraph is to provide some guidelines for rehabilitating coral reef habitats that have been impacted by oil spills. The rehabilitation process in the final step of a spill response effort that also includes actions to contain, cleanup, assess damages, and mitigate damages from a spill. Because our ability to rehabilitate damaged coral reefs is limited, this ordering of events also represents priorities for actions to be taken when a reef is threatened by oil pollution. In view of these priorities, the objectives of this chapter are threefold:

— To identify reef areas particularly sensitive to oil pollution incidents and to suggest means to minimize the threat of such incidents.

— To suggest techniques for assessing the impact of oil pollution incidents on coral reefs and their associated communities, and

— To provide guidelines for rehabilitating a reef after it has been impacted by oil pollution.

Coral reefs are structures created and maintained by the establishment and growth of populations of hermatypic corals and coraline algae. They may be comprised of wave resistant emergent of submerged reefal zones, or a combination of both. Geomorphically barrier and fringing reef protect the insular and continental coastlines that they border from erosion. Reef rock that forms terrestrial substratum on islands and recently uplifted coastal margins is often the sole or major component of limestone. Geologically older reef formations are found far inland and may serve as porous, oil bearing strata. Reefs typically provide habitat for a large variety of attached plants and epifauna, infauna mobile invertebrates, and fishes. The large numbers of economically important species they support make reefs locally important to commercial sport fisheries. The resultant high diversity and abundance of reef associates and the functional and spatial dominance of corals and coralline algae are the essential characteristics of coral reefs. Coral reefs are circumglobal in the tropics and subtropics between the norther and southern hemispheric 18.5 degree climatic isotherms. A majority of coral species and the most diverse and prolific reefs occur in Indo-West pacific seas. Tropical West Atlantic and Eastern Pacific reefs are generally less diverse in terms of caroals and reef associates, but are functionally quite as important.

In a recent interview of the impact of oil upon corals, Ray pointed out that extensive tanker traffic and oil-related activity occurs within the biogeographic zones occupied by coral reefs. Sufficient information exists to enable the production of detailed maps showing the location of coral reefs in relation to petroleum infrastructure. Such maps may be a useful aid in the identification of high risk areas. Two areas that have well developed reefs and much petroleum activity are given as examples in this discussion.

Having now made the effort to consider from theoretical point of view the system which it is thought may give rise to biological effects, it is no time to consider whether these effects can actually be observed under the conditions which prevail in the marine environment.

The marine environment is, however, subject to a wide range of influences the effects of which may swamp any effects which may be due to pollution. This is the well-known signal/noise ratio problem commonly encountered in the physical sciences. On the other hand, if studies are conducted in the laboratory the results can be no more than indicators of what will happen in the real world because it is quite impossible to reproduce the marine environment in the laboratory. Thus on the one hand we have the problem that the data obtained in what may be well-regulated laboratory experiments, in which the pollution parameter may be isolated for study, do not give results necessarily applicable to the marine environment, and on the other we have real environment studies in which the pollution effects are not isolatable and are usually swamped by other more influential effects.

The review also covers results obtained by exposing organisms to oil in the laboratory. Here again there are difficulties. Often the concentrations used in order to observe an effect are higher than could possibly be experienced in the marine environment itself or are continued for greater periods of time. Again, even when the concentrations are comparable, it is impossible to extrapolate to the real situation. It is quite clear that one can measure in the laboratory lethal and sub-lethal effects on organisms exposed to dissolved oil components and disoerbed droplets of oil in the water column. It is also clear that the concentrations of oil in the water column at real spills can be broadly comparable to those in the laboratory studies. In addition similar effects, both lethal and sub-lethal, on planktonic species can be observed at sea under real spills. When one attempts to evaluate the significance of those observations, however, in the only manner possible or meaningful, *viz,* by attempting to observe subsequent changes in fish populations, no scuh changes are discernible. There are a number of quite reasonable explanations for this state of affairs. Thus even if plankton populations were reduced significantly under oil slicks, phytoplankton is in continuous production and is rapidly restored as the oil alike moves on. Even if fish egg and larval stage numbers were similarly reduced it is fact that very few would have reached maturity anyway. The losses caused by the oil merely cause losses which would otherwise have taken place naturally.

The conclusions reached above regarding plankton in general do not of course rule out the possibility that larger marine organisms may be affected. When this possibility is considered, however, it is found that free-moving fish and crustacea are largely unaffected. Sedentary crestacea on the other hand may be subject to smothering with an oil layer deposited by the outgoing tide. Grazing molluscs and, of course, sea-weeds are liable to similar fate while filter-feeding molluscs can ingest oil in separating out food particles. Generally mortality is highest with fresh oil, weathered oil having little effect. In any case the effects are transient as regards populations and depuration occurs rapidly with filter feeders when exposed to clean water. Some species are not sufficiently mobile to escape, do not posses shells by which to isolate themselves or have other protective coatings. Worms living in sandy or muddy beaches may be killed if oil components penetrate the substratum. Of course, oil will not penetrate across the water-table interface and so this will occur mainly in well-drained beaches.

With birds it can be concluded that soluble and dispersed oil components will have no effect and we have now reached a pure example of the effects of physical coating. There is no doubt that this form of pollution has a very serious effect at least on individual birds, but here again the Royal Commission has not been able to demonstrate conclusively that there is any threat to populations. There is, however, considerable cause for concern, particularly in relation to major spills close to a breeding colony onshore. It is obvious that oiled birds are one of the main vehicles of public awareness of oil pollution and arouse widespread feelings of pity and revulsion.

There seems little possibility of affecting large marine mammals such as the cetaceans. Seals on the other hand may become oiled as they leave and reenter the sea and this could result in loss of insulation. It has been noted, however, that seal populations around the UK are generally stable and that grey seal numbers have been increasing dramatically for the past 50 years.

In the early 1970s attention was given to the possibility that polyaromatic hydrocarbons present in oil pollution might concentrate in the tissues of marine organisms and that those concentrations might increase up the food web to levels which could induce cancer in man. Research results now available, however, indicate that in common with other petroleum-derived hydrocarbons these compounds are rapidly released when the organisms are exposed to clean water and it is extremely unlikely that they could accumulate in the manner originally feared. It has also been shown that these compounds are in any case the inevitable products of the combustion of most types of organic material such as wood, garden rubbish and even tobacco, and as such are widely distributed in the environment. Furthermore they are present in many foods and it is therefore possible to show on the basis of dietary studies that the consumption of sea food would be unlikely to provide more than 2-3% of the normal intake. Recent research has also failed to establish a connection between this form of pollution and cancer in the marine animals themselves and it is extremely unlikely, to say the least, that ingestion of cancerous tissue from animals could cause cancer in man in any case.

The aims and objective of oil spill response should therefore be to protect amenity, fishery and wildlife resources and the efforts expended should ideally have some relationship to the quantification of threat to those resources if nothing is done. If nothing is done to respond to an oil slick far enough at sea for it to disperse through natural forces before it reaches shore, then minimum impact on the environment will result since there will be no amenity loss, and no effect on inshore fisheries. The oil will disperse far at sea when its effect is least. The only possibility of quantifiable damage resulting would be if high concentrations of seabirds became affected. If natural dispersion forces are unable to remove the oil from the sea surface before it reaches shore, then, in addition to seabirds, inshore fisheries, the flora and fauna of the inshore zone and amenity are at risk. In terms of quantified risk, therefore, there is no likely to be greater environmental impact than in the previous example. It is reasonable therefore to make greater efforts to respond in this case than in the previous one. It should however, be realised that efforts should be made to deal with the oil in the lower - risk are, i.e., at sea, before it reaches the comparatively high - risk area, i.e., the inshore zone. If it is possible to prevent oil reaching the high - risk zone, minimum damage will result. If response occurs only when the oil has reached the high-risk zone then the damage will already have occured and one is then no longer preventing damage but only attempting to assist the area to recover from the damage it has already sustained. It is therefore clearly preferable to deal with the oil in low risk areas before it reaches those of high risk. If this is impossible, then the aim should be to return the damaged areas to their original state as quickly as possible and the best way to assist in this will be to remove the oil as quickly as possible if natural processes are considered to be too slow.

There are three possible responses to oil spills at sea.

— It can be decided to leave the oil to be dispersed by natural processes.

— The natural dispersion rate may be increased by the use of chemical.

— Attempts may be made to remove the oil from the sea surface by mechanical means.

The nil response option will be possible when no resources are threatened and natural dispersion rates are fast enough to remove the oil from the surface before any resource comes under threat. The use of chemical dispersants should be seen as a means to increase natural dispersion rates so that resources are less likely to come under threat. It is of course preferable to remove the oil completely from the environment, and so mechanical recovery devices are favoured provided they will work sufficiently to deal with the oil in the low—risk zones before it reaches those of high risk. It is generally accepted that if recovery is into possible far at sea in a given situation, it may be preferable to remove the oil completely from the environment, and so mechanical recovery devices are favoured provided they will work sufficiently to deal with the oil in the low-risk zones before it reaches those of high risk. It is general accepted that if recovery is not possible far at sea in a given situation, it may be preferable to disperse the oil there and so prevent damage later in a high-risk sheltered zone than to wait and attempt to recover it when it has reached the high-risk sheltered zone. The latter option will not prevent damage. It will merely remove the oil after the damage is done. Such belted action will of course assist in the recovery of the area affected.

Clearly oil does not remain permanently on the sea surface. The pollutant may disperse into the water column either completely or partially under the action of natural forces and or it may come shore. If the oil is spilled close to shore there is little opportunity for natural dispersion before the slicks reach shore. In such circumstances beach pollution seems inevitable. However, even in the Amoco Cadiz it has been estimated that some 50% of the spillage dispersed into the water column in the prevailing conditions of high natural agitation rather than stranding on the beaches. With the advent of North Sea oil exploration and production, efforts were made to estimate the quantities likely to reach shore by considering natural dispersion rates in relation to the speed of travel on wind and tide. It had been concluded that for Ekofisk oil only very small percentages of the original spillages have any chance of polluting beaches. Thus although natural dispersion does take place, it cannot be relied upon in general to remove the pollutant from the sea surface before it reaches shore. The extent to which natural dispersior will assist is a function of oil type, quantity spilled and distance from shore of the spillage source.

It is important to recognise at this point that dispersants are simply as means of increasing the natural dispersion rate so that sufficient before it reaches the showe. Certainly it is necessary to treat an in which this is an advantage. Thus the spreading phenomenon itself is form of dispersion. In fact it is two-dimensional dispersion and all that remains to be done is to add the third dimension, i.e., to disperse the surface oil into the body of the sea. Oil concentrations measured under chemically dispersed oil slicks and under the naturally dispersing Ekofish Bravo oil slicks were much lower than 100 ppm because the oil, of course, disperses, throughout the sea water to even greater depths than 1m. Further experiments on this topic have been carried out on Kuwait crude oil at experimentally controlled thicknesses, with the fan-jet nozzle arrangement used for dispersant efficiency testing at sea. Here slicks up to 1 mm in thickness were produced in order to produce high initial concentrations in the water by use of appropriately enhanced dispersant discharge rates.

Maximum possible concentration in the top metre on dispersal of a 1 mm thick slick would be 1000 ppm as opposed to the 100 ppm arising from the normal 0.1 mm thick case. The aim of these experiments was to study rates of dilution of the dispersed oil into the sea following treatment. Wind speeds on this occasion were 8-10 knots with sea state 2-3. With this degree of agitation, observed concentrations were low, suggesting rapid transportation of oil to greater depths by natural turbulence as shown in Table 1. When these results are compared with the theoretical concentration of 1000 ppm. If all the oil remained uniformly dispersed in the top metre, there is a strong suggestion that initial concentrations in surface waters are reduced by dilution to greater depths. The results of these chemical dispersion experiments in calm and slightly rougher conditions suggest that while initial concentrations might be higher than for natural dispersion, subsequent dilution is rapid, even for slick thicknesses ten times thicker than those normally treated. As a result the persistence of significant amounts of oil in the surface waters is likely to be of shorter duration if dispersant are used than if they are not.

Table 7

Chemical Dispersion of 1 mm Thick Slicks of Kuwait Oil

Time after discharge (min)	Concentration of Kuwait crude oil (ppm) in upper metre of sea water		
	Run 1	Run 2	Run 3
0	34.3	24.2	0.85
1	—	15.8	—
2	47.8	—	8.7
2.5	—	12.2	—
5	—	9.4	—
7	17.8	—	3.5
10	—	5.2	—
15	—	—	1.7
18	1.9	—	—
25	—	4.2	—
40	0.8	—	1.35
50	—	1.9	—
80	—	—	1.5
100	2.2	0.8	—

Oil recovery at sea it is necessary to return to the earlier assessment of the problem. It will be recalled that the counter rate for a 1 m wide unit travelling at 1 m 0.2 tonnes/h. It is also a known fact that if a barrier having draught and free board a skimmer or a boom for example is moved across a water surface at speeds greater than 1 knot there is set up a flow of water beneath the barrier which is sufficient to carry with it any floating oil which is in front of the barrier: consequently the encounter rate in the oil recovery context cannot be increased by increase in speed of travel in the manner appropriate to dispersent application. At first sight the only option is to increase encounter rate by increasing the width of the collection unit by the use of booms. Thus a boom array of width 250 m will increase encounter rate of 90 tonnes/ h. Such an array, however, is extremely cumbersome in practice. To remove oil pollution from the marine environment before it could reach sensitive inshore areas and beaches and from a realisation that, in spite of claims to the contrary, no significant quantities of oil had ever been recovered at sea. The objectives of the programme were to find out why performance was so low and how to improve it with due regard to cost benefit concepts. To this end, available equipment was classified in terms of the nature of the skimming element, e.g. absorption belt, disc. repel vortex chamber of weir/direct suction device. In similar manner booms were classified in terms of their identifiable design elements. Examples of each type of stimmer and boom were then acquired from evaluation at sea in the presence of oil.

As a result of these investigations it has been found that recovery equipment fairs to perform satisfactorily in waves for the following reasons:

— The inertial mass of the equipment is too great.

— The linear dimensions of rigid elements, or of the entire systems, are too great in relation to encountered wavelengths.

This being so purpose—built vessels incorporating the skimming function in as a rigid integral element in their construction will be inoperative in waves. This conclusion means that great cost savings are immediately possible, since such vessels are extremely expensive to build end maintain and would if chosen be required in large numbers in order to overcome the enormous logistic difficulties inherent in this concept. Another system which can be reported on favourably is that developed by ORI/Star Offshore, known as the Force 7.4.

This is a single-ship system and is based on the flexible absorption rope principle. Consequently it has excellent wave—following characteristics and should recover oil of viscosities allowing adhension to the rope at rates controlled only by encounter rate. These two systems meet the identified requirements is different ways. They may different, however, in encounter rate potential, the Springsweep system having a collection boom while the Force 7 does not. On the other hand the Force 7 may be capable of operating at a greater ship speed. It is believed that further evaluation of these systems can now be achieved only through operation at real spills. Decisions can then be taken on a cost—effectiveness footing in the light or considerations present to here. The Eleni. V. incident involving a spill of heavy fuel oil which was solid at sea temperature, represented a particularly difficult case. The Antonia Gramsci incident, however, revealed that solid oil in the presence of ice presents even greater problems. In the end some 400 m of pure oil was recovered at sea (the actual recovery rate is not, available) and some 400-500 m^3 was subsequently recovered from 300 km of coastline at a cost of 17,500 m^3.

It has been shown that oil slicks on water are detectable by observing reflected light in the near—U.V. visible and near—IR regions of the electromagnetic spectrum. In addition, slicks may be observed in the thermal IR region because of apparent differences in surface temperature as between oil and water. In order to select the best wavelength windows for operation, however, it is necessary to consider the problem in greater detail. Apart from the surface—reflected radiance in which one is interest. There are two other background components contributing to the total radiance detected by a remote sensor. These are a volume—reflected component and a path radiance scattered by the intervening atmosphere into the sensor. This latter components depends on path length and at higher altitudes can be the predominant effect. In addition, the sources of the reflected radiance have to be considered, *viz.*, the sky radiance which is reflected at the surface component and the so-called global radiance which is the sum of the sky and sum radiances and is received by the sensor as the volume and atmospheric path radiance. Thermal radiance device measure apparent temperature which arises from the physical or 'real' temperature and the thermal emissivity. The presence of an oil slick can cause changes in both parameters. In the case of physical temperature an oil slick could in principle produce the following effects:

— The layer could absorb solar radiation and become hotter than the underlying and surrounding sea.

— Oil close to spill source could retain heat and thus be better than the sea.

— Evaporation of volatile components could have a cooling effect on the oil relative to the sea.

— Oil could reduce evaporation of water and thus produce a warming effect relative to the sea surface where water would continue to evaporate.

— Oil could restrict heat transfer between the atmosphere and the sea so that in warm weather the sea would be cooler under the oil and in cold weather it could be warmer than it would otherwise be.

— The movement of wind over the oil slick could warm it up by a mechanism of viscous dissipation of energy relative to the sea surface.

From the above list of possibilities it is clear that oil slicks at sea could be hotter than sea water or cooler, depending on the final outcome of a variety of processes individually giving rise to conffleting results, water. Oil is variously reported to have an emissivity 0.02 - 0.005 less than water, which may be expected to give a temperature contrast of 1-3 degree K.

This equipment is well worth investigating as a remote-sensing tool for oil on water. If satisfactory performance can be demonstrated one immediately has a number of advantages, via low cost and ease of operation together with real-time display of the imagery. LLLTV can have a spectral response in the 250-1200 nm region of the spectrum depending on the characteristics of the tube used. Oil has a greater surface reflectance than water in this region and so it should be possible to see oil in contrast to the water.

Some tank and wind tunnel studies have been carried out to investigate possible mechanisms to explain the difference in physical temperature between oil slicks and the surrounding and underlying water 5. In these experiments, thermocouples were places in the oil and in the water immediately below the surface at 1 mm depth and the bulk water at a few centimeters depth. Control tanks containing only water were similarly equipped with thermocouples at 1 mm and at a few centimeters depth.

Ultraviolet and infrared line scanners, have received considerable attention in a number of countries and the Daedalus system features prominently in this work. Canadian and US workers have evaluated the Daedalus 1230 dual—channel line scanner which is fitted with a UV detector sensitive in the 270-370 nm region and a thermal IR detector for the range 850-1250 nm and the Daedalus 1260 multi-spectral scanner in which the near — UV to near — IR region, 380-1100 nm, is divided into ten channels and there is in addition the usual thermal IR part at 850-1250 nm

The UV imagery from the dual-channel unit gave good contrast for the thin regions of the slick. This device had a better performance than the multi-spectral scanner mainly because of the lower noise equivalent radiance value for the UV detector. This fact indicates that features other those such as detector sensitivity and noise levels will affect results and to some extent therefore the results obtained apply only to the system used.

Microwaves sensors are available with wavelengths in the range 0.1-100 cm and can operate either in the passive mode, measuring difference in brightness temperature between oil and water at a given wavelength, or in the active mode as radars receiving a back scatter from the sea or oil surface.

Waves at sea are of two types, long wavelength, large amplitude gravity waves and small capillary waves, the latter being the first to be produced as wind speed increases from zero and which ultimately are super imposed on the larger gravity waves. Radar backscatter is mainly due to these capillary waves, but imaging radars can see gravity waves because of the increased concentrations of capillary waves on the downwind faces of gravity waves as compared to the upwind sides. This produces a periodicity in the radar backsctter of capillary waves even though the radar wavelength is very much smaller than that of the gravity waves. This is turn suggests that backscatter intensity will vary with viewing direction with respect to wind direction.

Microwave radiometry is a passive technique which selects changes in microwave brightness of the sea in the presence of oil slicks in a manner analogous to the infrared technique. It also offers the possibility of measuring oil slick thickness with the added advantage over infrared that it is independent of atmospheric water content i.e. fog or rain. Calculations of microwave brightness of oil over water for a given frequency predict alternating maxima and minima as the oil thickness increases. By using two frequencies, ambiguities can be removed and thickness uniquely determined over a wide range. Measured antenna temperature is the average of the brightnes temperature over all directions weighted by the antenna response pattern. The total brightness temperature in any direction is due to the radiation emitted by the surface, the downward radiation of the sky reflected by the surface, and the emission and attenuation of the atmosphere between the surface and the radiometer. The quantification of oil in effluent waters will be discussed. In both these situations the source of the oil is known and it is the quantity of hydrocarbons in the water which is being sought. In the aspect now under discussion it is the source of pollution which is to be established. This requires a different approach. Before discussing these different requirements, however, it may be worthwhile to recall whoes analysis may be necessary in this context.

Aerial surveillance with standard photographic equipment is not in itself sufficient to establish that a named ship identified in a photograph, is actually responsible for discharging the oil which appears to emanate from that ship in the photographic evidence. In fact the best way to establish that the vessel is discharging oil is to sample the area which shows as an oil slick in the photographs and to characterise that sample by chemical means in such a way as to show that it is identical to a sample of the oil carried on board. It is worth recalling also that this comparison between pollutant and putative source may be advisable even in the case of a shipping casualty where attribution of oil source might be thought obvious. This can rise when passing vessels take advantage of the widespread pollution arising from a casualty in a given area to discharge tank washings, bilges, etc. As a result some of the oil arriving on shore may be erroneously attributed to the casualty and the operational discharge offenders may escape detection.

It has already been stated that the best way of establishing that a pollution sample originated from a suspect ship is to compare the results of chemical analysis of the pollutant with that of the oil carried by the ship. In fact it is the only way. Such is the complexcity of individual oils and the wide variety of potential sources of distinct oils that it is quite impossible on the basis of a pollutant sample to state the source of that pollutant. Generally speaking, the best that can be done with an unreferenced sample is to say that it is a fresh crude oil, a weather crude oil, a light or heavy fuel oil or tank washings. This is quite insufficient to associate it with a particular source. This being so, it is necessary in a case of suspected operational discharge to sample all the oils on board including bunker fuel and lubricating oil. This may involve sampling all the tanks, though reference to the cargo manifest may help to reduce the amount of required sampling. Similarly in a shipping casualty it is useful to sample all oils on board since damage to the vessel can result in the release of an variety of oils.

In collecting samples of liquid oil from the surface of the sea, scare is necessary to ensure that oil is indeed being collected and that the result is not simply a sample of dirty water in an oil contaminated container. Where the pollutant is a relatively thick layer of water-in-oil emulsion it is easier than if the pollutant is fresh oil present to about 0.1 mm thick or even less, but in all cases it is easier to collect water than oil. The use of a container with a bottom outlet and tap is helpful in collecting sufficient oil since with such a device water can be drained off and sampling repeated if necessary. A funnel can be used in this way, using the finger to release the water and retained the oil. Alternatively where the oil layer is very thin or is in the form of many small isolated globules of oil or emulsion, it may be possible to use adsorption pads. These may be of Teflorn, steel or polyurethane. The oil sticks to the pad when it is placed lightly on the oil-contaminated sea surface. The pad is then placed in a sealed air tight container for transport to the laboratory. Oil can be removed from the pad by use of a suitable solvent and this operation can be carried out either before transportation to the laboratory or after arrival there.

Everything that has been said in the chapters on at-sea oil spill response holds also for inshore waters. However, there are a number of additional points of specific relevance to the inshore situation. As oil approaches the shore, the efforts made to combat it either by dispersants or by mechanical recovery means will of necessity be conducted closer and closer to shore, and if these efforts are inadequate the oil will eventually strand on the shoreline. The extent to which the oil stands will depend on the extent to which it is treatable at sea and this in turn will depend on soil size, oil type and the proximity to shore of the original spillage. So far, beach-cleaning costs per tonne are considered to be so high as to justify operations at sea.

For the moment it will be assumed that if activities at sea were justified, they presumably still are close to shore. The next step is to consider dispersants and mechanical recovery separately with a view to establishing whether their ought to be in reasons for discontinuing activities commenced at sea simply because the oil is coming closer and closer to shore, and alternatively the extent to which it might actually be easier to carry out the desired operations now that the oil slicks may be reaching more sheltered waters.

In cases of oils which are not amenable to dispersants, only recovery is possible. As with dispersants, the establishment of water in-oil emulsions with their increased viscosity will render the recovery of oil more difficult. On the other hand, the possibility of thicker layers brought about by the booming effect of the beach itself (the best boom available) may provide greater layer thicknesses and therefore the possibility of increased encounter rates for oil recovery systems. In addition, unless the inshore waters are exposed to the full force of the weather the general sheltered nature of many inshore situations will assist in oil recovery operations. In general, therefore, provided there is enough sea room and depth of water for ship manoeures, the deep-sea equipment will be usable in sheltered inshore conditions. However, in addition, because of the shallower water depths it may be convenient to more booms rather than to tow them. Moore of booms, being static of course, do not increase encounter rate but are used to protect specified stretches of shoreline and/or to deflect moving oil to a slected recovery point. These techniques are specific to the inshore situation.

In general, mechanical recovery of oil and water in oil emulsions should be easier in sheltered waters than at sea. Indeed it may be better to view oil recovery as a sheltered water technique which may in certain circumstances by usable at sea rather than to present it as a sea going technique which can automatically be used in sheltered environments. Be that as it may, boom utilisation and oil recovery techniques in relation to the particular inshore situation. Where sea room and depth of water limited such operations there is scope for smaller devices deployable from smaller vessel, e.g., the net system from inshore fishing boats. However, the use of small vessels is not without its difficulties. The logistic problem arising from lack of storage on board mitigates against the small vessel approach. Again, if the oil is already against a harbour wall or other coastal structure it may be simpler to deploy the oil recovery equipment directly from the land. There is little point in working from a small boat if one can work directly from the river bank or quay wall or shore. Recovered fluid is ther already ashore and one step in the logistics of water-borne operations is immediately eliminated. In considering how best to deal with oil once it has come ashore it is necessary to have regard to the type of beach material affected.

Beach material is classified in terms of grain size. Mud is the term used to describe grain sizes less then 0.05 mm. Mud beaches generally have a slope of less than 1 degree they develop in areas of low water velocity which permits the deposition and retention of such fine material. Such beaches are saturated with water at all times and because of the low drainage which this implies, the surface is usually covered with a thin layer of water. They have very low load-bearing capacity and are generally speaking of high biological value.

Sand beaches have a particle size range from 0.05 to 2.0 mm with beach slopes between 1 degree and 40 degree. Such beaches are generally more exposed than mud beaches and are subject to seasonal erosion and deposition cycles depending on the energy of the incoming waves. Drainage rates are higher than for mud but again permanent saturation and surface water layers may be encountered.

Load-bearing capacity is higher than for muds, Sand beaches are of low biological value. Gravel and pebble beaches are of particle size 2.0-50 mm and 50-256 mm respectively and tend to be narrower and steeper as the particle size increases from gravel to pebble. Such beaches are again of low biological value. Apart from these physical difference due to particle size, other parameters are relevant such as exposure of the beach in question, the onshore energy levels of the waves to which it is exposed and the accessibility of the beach from the point of deploying pollution response equipment.

It is important to distinguish the specific uses of booms. In the first place the booms may be used in a sweeping mode to concentrate oil towards a skimmer so as to increase the oil encounter rate of that skimmer. In the second place booms may be used to protect certain coastal resources from oil contamination or to deflect oil on to a nominated or sacrificial beach for recovery. In the former case the booms are deployed from ships and are used in a mobile mode. In the latter case they may be, therefore used in a stationary mode. Of course many aspects of boom design are common to both applications, but it is as well to be aware of those aspects in which boom design is specific to one situation or the other. Some booms are usable in both the towards and moored conditions but others are not. It will rarely be possible to deploy moored booms in such a way as to protect the very large lengths of open coastline which may be threatened by oil pollution in a major incident. The use of such booms must be selective. In addition, depending on time-scales, logistic problems, etc., it may be impossible to have such booms in place in the required time. However, in some cases it will be possible to consider these matters as part of an overall plan for cargo transfer in a safe haven. In such a case the deployment of any necessary booms will require to be completed in good time for the arrival of the casualty.

8 Effects and Control of Marine Pollution

In this chapter we shall attempt to get some idea as to what pollution does and why we consider it to be so undesirable. There are three basic type of pollutants: the pathogenic, the aesthetic, and the ecomorphic. Pathogenic pollutants are those which cause disease. This disease may be fatal if the pollutant is a lethal poison. Aesthetic pollutants are those pollutants causing a change in the environment displeasing to the eye, ear, or nose of man. Ecomorphic pollutants, on the other hand, are those pollutants which produce a change in the physical characteristics of the environment in such a way that there may be drastic changes in the structure or composition of the biosphere. Obviously, these three types of pollutants are not of equal import.

Pathogenic pollutants are certainly much more serious than the other two; however, ecomorphic pollutants are often an indicator of more serious types of pollution to follow . Furthermore, since pollution is man-produced and the effects are suffered by man, aesthetic problems are certainly of interest. Nevertheless in any management situation when trade-offs are required, it will certainly be necessary to rank different types of pollutants so that society can determine how it would like to spend is limited funds of time, effort, and money.

Acuate Effects

Let us look first at the pathogenic pollutants since these are obviously the most important. The effects of pathogenic pollutants are either acute or chronic. Acute effects are those which occur in a relatively short period of time while chronic effects may take time to be noticed, as in the case of carcinogenic pollutants, or the effect may be noticed immediately but continued for a very long period of time.

The acute effects are the easiest to determine because of the short time between administration and affection. The easiest way of measuring acute effect is by feeding the pollutant in question to test organisms, such as since or selected fish, and observing the dosage required to will 50% of the organisms involved.

This is usually done by feeding a pollution of test organisms increasing amounts of the pollutant, and for each dosage the number of individuals that succumb is noted. When these data are plotted, an S- shaped curve results, wherein dosage versus number of mortalities is plotted. From this curve it may be seen that there are a small number of individuals that require an extremely large does before they are killed while some are killed with a relatively small dose.

Another way of examining acute effect is to observe the effect of a given pollutant concentration on organisms over a somewhat longer period of time, noting how long it takes for given dosage fo produce mortalities. This say be done by giving a predetermined concentration of the pollutant to the test animals and observing how long it takes to kill half of them.

When data from an experiment such as this are plotted, a curve similar to that shown in graph 3.2 results. As may be seen, the curve is a hyperbola with the horizontal asymptotic suggesting a threshold concentration. In other words, below this asymptotic value there is no apparent effect, since at this concentration an finite amount of time is required for mortality. The vertical asymptote, on the other hand, appears to be a saturation level. At this level any increase in the concentration will not produce any further decrease in the amount of time required for fatality. Evidently there is a finite time required for the pathogen to do its work, so that no matter how high the concentration, the time required to kill will not go below this amount.

Charonic Effects

With the asympototic values, suggests another concept : the threshold value. There appears to be a dose size for each pathogen below which the pollutant has no effect. The concern with threshold values for various pollutants is one aspect of the study of chronic effects. Chronic effects are the most difficult to measure because of the time involved; in some cases cancers have developed as long as 25 years after initial exposure to the carcinogenic agent. Neverthless, measurements have been made and for many materials threshold value seems to describe the data reasonably well. Most experiments need some measure of the effect on the organisms other than death, so that many different measures are used. These arc : deformity in the growth of organisms, damage to particular organs, change in heartbeat or breathing rate, genetic damage to individuals, change in the rate of increase of a population, life expectancy of individuals within a particular population, and fecundity of females with the population.

Unfortunately, and index may be very descriptive of the effect of a pollutant on a particular species, but it may be completely ineffective in describing what happens when this pollutant comes in contact with other species.

A typical experiment using one of these measures would involve the variation in pathogen concentration while noting the variation in the heartbeat or some other measure. When these data are plotted the result is very often a straight line as shown in graph 3.3.

This linear relationship between effect and concentration may be extrapolated back to determine the concentration for zero effect. When this is done the line will cross the axis at a concentration from zero. This is called the threshold value.

A great deal of caution must be used in interpreting these data. Some pathogens apparently have no threshold value associated with them, and it is always risky business to extrapolate a straight line. The actual manner in which the organisms reacts to the stimulus might very well be nonlinear at lower dosages. In addition some chronic effects may be so suitable that two or three generations of the organisms might be required to observe any change.

Synergism

Another complication in the examination and quantification of pathogenic pollutants is the phenomenon of synergism. A synergistic effect occurs when the combined effect of two or more materials acting together is greater than would be expected from the simple sum of the individual effects. Temperature, for example, is a very strong synergistic parameter, generally enhancing most pollutant effects as temperature is increased.

The same is true of salinity and dissolved oxygen. Some pollutants, on the other hand, tend to decrease the effects of others, as might be in the case of a strong acid and a strong base present at the same time. Other pollutants may cause strong pathogens to be precipitated out of the water column or to be collected on sediments, so that in either case they end up on the bottom. Thus it is extremely important in the analysis of any pollutant to be aware of the other materials present and their synergistic tendencies.

Within the marine biosphere many organisms are affected by various pollutants. Whether or not a particular organism is affected by any given pollutants will depend on any variable. For example, many polluants are found only in he water column while others, such as some of the pesticides and heavy metals, are adsorbed onto suspended sediments and usually find their way to the bottom within a relatively short period of time. Consequently, these latter types would tend to affect those organisms living on the bottom or feeding from bottom organisms more then they would affect the organisms living within the water column, such as free swimming fish.

Also of import in determining the effectiveness of a particular pollutant on living organisms is its relative solubility in water and oil. Pollutants more soluble in oil will tend to find their way to organisms that have a larger content of oil in their body issues. This is particularly true of some of the pesticides which are more soluble in oil than water. The individual habits of marine organisms also affect their susceptibility to a particular pollutant.

Feeding habits with respect to time of day, portion of water column for aged, type of material ingested, and the method of digestion utilized all determine to a great extent the types of pollutant the organism will be exposed to. Similarly spawning habits determine the type of area and time of year in which a species is particularly sensitive to subtle changes in the environment. These changes affect not only the parents but also the offspring, and may be even their progeny.

Other habits not having to do with spawning also help to determine whether an organism will be exposed to particular pollutants. Weather the organism is sensitive to light or sound, for example, will often determine whether it is attracted to a particular out-fall. Thus we find that a pure and simple cause and effect relationship between a pollutant and an organism is not enough to determine completely the actual response of the organism in the real world. Laboratory results might be completely refused in the field simply because the organism does not respond to extraneous characteristics of the environment under natural conditions in the same manner in which it did in the laboratory.

Once some of the pollutant has been ingested by marine organisms, the major concern of man is the possibility of this pollutant appearing on his dinner table and causing a pathogenic response. Obviously if a fish ingests a pollutant and a human eats that fish, then that person will have ingested a does of that pollutant and there will probably be some danger. The question is, "How much?" In order to arrive at an approximation of the actual amount of undesired material finding its wax into the reader's stomach, one mush follow the pollutant through a number of steps.

Most organisms living in the sea tend to concentrate materials existing in the sea to much greater values that they are in the ocean itself. For example, a diatom, a small form of marine algae having a silicate frustule, has a body concentration of silicon about 40 thousand times as great as the oceanic waters from which the organism derives all of its material. Thus the diatom concentrates silicon very effectively so that if we eat diatoms, a large portion of our diet would be glassy. Similarly, other organisms, including fish, will concentrate other materials even if these do not naturally occur in the environment. But this amplification or concentration does not stop with one step. We must continually keep in mind the fact that an organism does not exist by itself in the marine environment.

Any organism is dependent upon many other organisms for its existence. An edible fish, for example, might very well subsist on smaller fish which, in turn, might subsist on small zooplankton, which, in turn, may subsist on phytoplankton. There may be 5 to 10 individual links in a food chain leading to man, and there is the distinct possibility that for each one of these links a concentration of the pollutant occurs.

Thus the total pathway to man for the pollutant might be a rather tortuous one, but it might result in a very high concentration of the pollutant material in the fish. This in itself may be of no concern to the individual unless he ingests the fish and if so, how often and how much he consumes. It is not enough simply to know that there is some pollutant present in marine organisms : one also should know how much is present, and in addition two other facts are required.

This first is the accepted maximum level of ingestion of the pollutant before harm results, and the second is the amount of fish that can be eaten before this level is reached. In this way it might very well be possible to have a particular kind of fish one or two times a year with perfect safety, while if it is eaten once a day, it might very well have a toxic effect.

Ecomorphic Pollutants

There are many different types of pollutants that may be classified as ecomorphic, while some pollutants fall into all three of the classes stipulated above and many fall into at least two. Sediment is a pollutant whose effect is primarily that of changing the environment since it is not pathogenic and very often is not aesthetically displeasing, although it can be. As it settles to the bottom, it will tend to bury the organisms that live on the bottom, such as cysters and clams, and eventually smother them. Sediment also tends to fill in marsh areas, killing the marsh grass, and completely changing the habitat. Since marsh areas are breeding grounds for small fish and an important ecological link in most of the oceanic life, any destructive process such as this is bound to markedly change the marine biological environment.

On the other hand, in some cases where man's activities have produced large amounts of erosion and sediment has been brought down by rivers into estuaries, some estuarine areas have been filled enough to support marsh grasses. In some cases excess erosion can produce marsh lands rather than destroy them. Whenever there is a depth change in an estuarine or coastal area, the effects are many.

They range from changing the biological environment so that larger fish will no longer live in the shol areas to affecting the total circulation pattern of the area. But suffice it to say that changing depth in many cases does change the manner in which salt and fresh water mix in the coastal area and also the amount of flushing present in a particular region.

When the sediment is in suspension it also has an effect on the eco-system, the most obvious aspect being a change in transparency. By making the water more opaque, light will not penetrate as deeply as before so that the thickness of the layer of water in which plants grow will be decreased. If the water is fairly shallow to begin with, this decrease in transparency may very well result in the inability of rooted plants to grow and some marsh lands may be destroyed.

The effect of sediment, then is two-fold : changing both bottom and water properties. Both of these effects tend to change the physical characteristics of the ecosystem.

Another ecomorphic pollutant is oil, although oil sometimes produces toxic effects. The result of oil coating bird feathers is a well known physical effect. The oil does not poison birds; it simply makes it impossible for them to fly and consequently many of them die. A film of oil on the surface will also decrease the amount of sunlight entering the water and, similarly, will limit the exchange of oxygen from the atmosphere. In addition to the aesthetic and pathogenic effects of oil, there are thus also ecomporphic effects.

The toxic effects of various nitrogen compounds are well known but fertilizers and other nutrients washed into the sea have a nontoxic effect on marine waters. These materials will do just what they are designed to do : make plants grow, whether these plants be terrestial or aquatic plants.

The toxic effects of various nitrogen compounds are well known but fertilizers and other nutrients washed into the sea have a nontoxic effect on marine waters. These materials will do just what they are designed to do : make plants grow, whether these plants be terrestial or aquatic plants.

Those plants on the bottom of the blanket will tie, due to lack of light, and those roting plants will utilize oxygen in their decay processes. Even though there are more plants than before, producing more oxygen in the process of photosynthesis, so many of them are dying and decaying the result is to decrease the dissolved oxygen. This phenomenon is called "Eutrophication." Another ecomorphic pollutant that seems to be receiving a lot of attention lately is heat. Usually supplied to the marine environment form electrical power plants or other industrial sources, this heat will increase the water temperature and may result in different kinds of damage to marine organisms. Since most mutations are caused to thermal activity under normal conditions, the possibility of genetic damage is real. It is estimated that about 90% of the mutations occuring normally are thermally caused, while only about 10% are caused by radiation.

Aesthetic Pollutants

The last type of pollutant mentioned above was aesthetic pollutants. Aesthetic pollutants are those which offend the human senses and as such are considered by many to be unimportant. In many environmental problem areas, though, the aesthetic considerations often receive more attention by environmentally oriented citizens than some of the pathogenic pollutants. This is simply because the aesthetic pollutants are visible and what can be seen, smelled, or heard appears much more bothersome than that which is hidden, even though the hidden pollutant might be much more dangerous to life.

Consequently, any scheme for controlling marine pollution must include some consideration of the aesthetic pollutants if it is to succeed. An excellent strategy seems to be to focus on the aesthetic pollutants and use them as a machanism to gather momentum to clean up the rest of the pollution problems.

Power transmission cable has caused more than one power company grief because of the environmental consideration given the local citizens who are concerned about the appearance of power cables running across the countryside. Oil wells, especially the offshore variety visible from resort areas, are of great concern to most environmentalists.

The same thing is true of equipment required for dredging operations and the spoil areas resulting from these operations. To some people, especially those living in relatively pristine environments near the water, boats can also be an aesthetic pollutant, especially those that cause waves and noise.

Effluents that cause the colour of the sea to change to increase the amount of suspended sediments also may be considered aesthetic pollutants. It simply is not as pleasant to go swimming in water that is cloudy and contains colouring material different from what one would expect in a normal, natural body of water.

In essence, any change from the pristine environment appears to many to be aesthetic pollution. Obviously we are living in an unreal world if we think the marine environment can be maintained in an unspoiled form. It should be borne in mind that many complaints about aesthetic pollution come from individuals who would father live in the primeval environment not realizing that it would not be available to them unless the commercial world about them supported it.

There appears to be a dichotomy wherein try to maintain the unspoiled environment in certain areas while letting others become over utilized. Nevertheless, there is a growing movement to include in the cost of every project a relatively small sum to improve its aesthetic qualities. This is, of course, exactly the same philosophy that is applied to the control of other forms of pollution.

Additional money is spent in a chemical process, for example, to control the pollution-producing properties of the effluent since society believes that this additional sum of money is desirable expenditure. It may similarly be maintained that the expenditure for aesthetic pollution control is desirable as it also will tend to increase the quality of life. This paragraph briefly state those portions of these studies that are relevant to determining the effects of oil on marine wetlands. Also, the literature and out personal experiences with oiled marine wetlands are synthesized to allow an evaluation of methods or protection, cleanup, and restoration attempts that have been carried out in marine wetlands. This section accomplishes there two objectives by:

— Providing a brief review of the effects of all spills and related celanup activities on salt marshes and mangrove ecosystems

— Reviewing methods of protecting marine wetlands from being oiled.

— Reviewing successful means of cleaning marine wetlands following oil spills.

— Reviewing and presenting techniques that have proven successful in restoring marine wetlands damaged by oil spill and/or cleanup operations.

— Establishing a set of criteria and discussing guidelines for decisions on means of protecting susceptible areas, and for cleaning and restoring oiled marine wetlands.

— The differences in growth patterns and life cycles are essential elements in consideration of methods of cleanup and restoration of marine wetlands.

— It should be noted that salt marsh and mangrove vegetation occur together in some parts of the world, e.g., southern Brazil, southeast Australia. For the purpose of this discussion, we need to identify two terms.

— Recovery : return of a site to dominance by native organisms that are within the natural range of limits for structure and function in unprotected examples of the ecosystem within the local geographic area.

— Restoration: man's efforts to initiate and/or enhance the recovery process.

Effects of Oil Spills on Marshes and Mangroves

Oil impacts to marshes will vary with a number of factors including the amount of oil, type and magnitude of cleanup activities, type of oil, physical and biological structure of the marsh, latitude, and season. Subsequent persistence of damages and recovery of the system also depend on numerous physical, chemical, and biotic factors, including weathering rate and degree of removal or retention of oil, availability of propagules (seeds or vegetative), successional processes, sediment erosion/ acceration, and restrorative activities by man. Plants can be affected in several ways. Physical smothering by oil can lead to reduced transpiration, respiration, and photosynthesis. Absorption of toxic oil fractions through the leaves or roots may cause poisoning of the plant by disrupting cell membranes and cellular organelles. Some species such as *juncusmaritimus is gerarii,* and Spartina anglica exhibit epidermal or cellular resistance to oilds.

There is a considerable range of sensitivity to oil. Some species are resistant even to repeated spills (e.g., salicornea sp. Many of the more resistant species are also pernnials with large root systems that enable them to grow back following oil damage to aerial portions. Oils vary in their toxicity, according to the content of lowboiling compounds, unsaturated compounds, and aromatics. In general, the higher the concentration of these constituents, the more toxic the oil. It is self evident that the amount of oil and frequency of oiling will effect the degree of damage, but it is less obvious that viscosity of the oil is important.

With light oil, there is a physical limit to the amount that will "stick" to a shore, with subsequent tides redistributing oils. In contrast, the viscosity of heavy oils often allows retention of large volumes and many cause mortality not through toxicity but through smothering. Heavy oils also tend to be more persistent and to extend the duration of damage, thus slowing recovery.

Since and sediment structure of the marsh considerably affect the degree of damage and potentials for recovery. In the United States, Europe, and elsewhere marshes may be narrow fringing bands along the high water mark, especially in areas where tides are small, or marshes may be broad vegetational belts (occasionally many hundreds of meters wide) in estuaries or on sheltered coasts, especially where tidal ranges are large.

Extensive marshes may not be completely covered even to massive spills, as oil will tend to affect the sides of the creeks and drainage channels first and then progress into the marshes plateau. Where vegetation it tall and dense (e.g., in a spartinetum of juncetum), considerable volumes of all may be completely oiled, and after severe damage, there may be no source of propagules for recovery. In such area, restorative action may be the only means of marsh reestablishment in the short term.

Sediment characteristics, including particle size distribution and drainage patterns, considerably affect potential for oil retention. Marshes in North America and Europecan be found in sediments ranging from medium grainted sands to very fine silts where more than 95% of sediment particles are less than 63 in diameter. Sands tend to be more freely drainging and more easily penetrated by oils as shoots can occur.

If oil penetrates anaerobic sediments weathering is reduced or stops, and the oil may be retained in unchanged form for considerable periods of time, reducing or preventing plant regrowth. However, where oil is weathered before or after incorporation, even very high levels may not prevent plant regrowth. Growth in solicornia sp., *suaeda maritima, Aster tripolium* and *Spartina anglica* has been reported run areas where levels of total aliphatic hydrocarbons of up to 5,875 µg/g were found.

Seasonal changes in the marsh may considerably affect the extent of oil damage. Oiling before or during flowering can cause reduced flowering and seed production. This may delay flowering and seed production (Dicks, unpublished data). During winter months many species die back and oil may affect only dead and decaying vegetation; however, in spring and summer growth, new shoots and germinating seedlings are particularly vulnerable.

In some cases, a major problem following oil damage to vegetation has been loss of sediment stability followed by erosion (e.g., Isle Grande, France following the Amoco Cadiz oil spill). Where extensive rhizomatous roots mats ramify the sediments or where vegetation cover is continuous (e.g., in denise Spartina Stands), sediments may remain stable even when vegetation has been killed. In low marsh colonized mainly by annual species (e.g., Salicornica) or where vegetation is sparse, root mats may be insufficient to prevent crosion.

Where erosion occurs, lowering of marsh level may prevent reestablishment of vegetation. Erosion may also follow damage to creek systems after cleanup activities (e.g., during the Amoco Cadiz spill that affected the Isle Grande marshes. The importance of slat marshes in prevention of coastline erosion is well established and may be of prime importance in some areas.

Because of the preceding considerations, basin forest may have variable characteristics. Generally, however, basin forests are well developed with canopy heights that may reach 15-20 m.

Riverine forests: This forest type develops along the edges of estuaries, often as far as the inland extent of the saline intrusion. It is best developed, however, in the lower course of the estuary where water flows are moderate to high and high nutrient inputs prevail. This combination of adequate freshwater and nutrient inputs allows the development luxriant stands.

On the periphery of the forest, the dominant species is the Rhizophora mangrove. This is an area where there is greater kinetic energy by due to the combination of river flow and tidal motion. The complex root system of Rhizophora, with its extraordinarily developed adventitious ("stilt") roots, allows the establishment of large trees on very unstable, soft soild.

In the inner forest, mixed stands of Laguncularia and Avicennia occur. Rieverine forests are highly productive with litter fall rates in excess of 4 g/m/day. Canopy heights reach 20 m in the best-developed forests.

Fringe forests: This type occurs along the seaward edge of forests lining protected shorelines. There is usually a strong horizontal gradient in tropography as well as a significant gradient in turbulence and tidal amplitude. Energy derived from the tides and waves decays rapidly inland where tidal flooding is further reduced by the rise in the mangrove floor.

The reduced flusing inland results in the establishment of a physiochemical gradient in water quality, toward the outer edge of the fringe, where greater energy levels prevail, the large, heavy seedlings of Rhizophora become established. Inland this way and tidal energy dissipates, and in the higher, less flushed parts of the fringe, Avicennia becomes the dominant species. Since frings are exposed, they are subjected to periodic destruction buy storms, waves, or scouring by strong current as. During storms large amounts of debris may be deposited in the outer firinge, reducing circulation to the inner firinge and the basins inland.

Overwash forests: These forests are formed on small island or projections from land masses (peninsulas) they are characterized by intense flushing by daily tides. The high flushing rates do not allow the formation of strong salinity gradients nor the accumulation and aggradation of the soil or of nutrients factors that usually lead to a succession of species toward the interior. True overwash forests are thus monospecific (cominated by rhizophora).

The amount of flushing and nutrient availability are critical factors that determine the structure and the rate of overwash forests. They, overwash islands in coastal oligotrophic waters may not be as developed as those inside coastal lagoons where higher nutrient accumulation is possible.

Excessive flushing leads to erosion and loss of the forest whereas extremely low flushing rates lead to accumulation of debris and vertical aggradation inside the forest with reduced flushing rates toward the core of the island. These islands thus evolve into firings and basin forests. Cintron *et al.* have described the formation, ageing, and decay of these island in southwest Puerto Rico.

Dwarf forests: Dwarf forest are found in marginal environments where their structural development is limited by edaphic factors, probably the unavailability of nutrients, Mature trees in these forests are usually less than 3 m tall. Salinity is not the limiting factor. Although the vegetation is limited in high, leaf size is not reduced as usually happens when mangroves are under the influence of a chronic stresson. This type of forest is different from those that may be considered scrub forests and that develop under the influence of chronic stressor, such as high salinity or cold stress. The dominant species in Dwarf forest in Florida and Puerto Rico is Rhizophora.

Worldwide, where some documentation of impact to mangroves has been observed and reported. Most of the earlier reports are anecdotal with little or no quantification as to the levels of oil contamination; the actual types of oil stranded in the forests; the areal extent of mangroves oiled, and actual numbers, sizes, or areal coverage of magroves killed or stressed by the oil spill. In addition, the species of mangroves involved was often not mentioned.

In addition to impacts to the trees themselves, the final component of the ecosystem is usually treated by noting dead or stressed animals at the time of the spill, or attempting to quantify losses or changes by one-shot sampling. In particular, the sampling of infauna, in many of these cases, has not taken into account the normal seasonal variations in infaunal species and numbers. All these problems may lead to widely different observations of the "impact" of a spill by different investigators.

Due to the fact that floating oil is transported by waves and currents and strands on shorelines, low-wave energy ecosystems such as mangroves forests are routinely sites where oil accumulates after a spill. In addition, the inaccessibility of most mangrove forests makes oil removal very difficult, if not impossible, and the fine-grained and peaty anaerobic sediments characteristic of mangrove forests may reduce the rate of microbial breakdown of oil.

These factors, plus the burrowing activities of crustaceans characteristic of mangrove forests, can lead to persistent, high levels of oil contamination not only on the soil's surface but also deep in the sediments in the mangrove root core, and mangroves can be killed.

Table 1 lists generalized responses of mangroves to oil spills in two general categories—acute and chronic exposure would be short term (1.4 weeks) and may and with no further impact if the oil is removed either mechanically or by natural wave and current activity.

Table 1

Generalized Responces of Mangrove Forests to Oil Spills

Stage	Observed Impact
0-15 days	Death of birds, tutles, fish, and invertebrates
15-30 days	Defoliation and death of small [<1 m] mangroves-loss of aerial roots
30 days-1 year	Defoliation and death of medium [< 3 m] mangroves-tissue damage to aerial roots
1-5 years	Death of large [< 3 m] mangroves, loss of old aerial roots and regrowth of new ones (sometimes deformed)
	Recolonization of oil damages areas by new seedlings
1-10 years	Reduction in litter fall, reduced reproduction, and reduced survival of seedlings
	Death or reduced growth of young trees colonizing spill site?
	Increased insect damage?
10-50 years	Complete recovery

There are many factors, acting both individually and in combination, which govern the effect of oil on marine life. These will include the type and quantity of oil, the duration and quantity of pollution experienced by the organism, the state of the oil, e.g., fresh, weathered etc., the season, with respect to the annual cycle of the organism, and the habitat and the natural stresses to which the organism is subjected.

Any or all of these can affect the organism in many different ways. The effect may be lethal or sub-lethal, in the latter it can disrupt physiological or behavioural activities and death may result due to interference with feeding of reproduction. The uptake of oil can result in tainting or possible carcinogenesis either of the organism contaminated or one higher up the food chain.

The visible and well-known effects are beach pollution and the death, due to oiling, of seabirds. It is not unreasonable to wonder, if not actually to fear, whether oil might nor adversely affect marine life and, in particular, harm fish the source of much food for man and his domestic animals either by reducing their numbers of rendering them unfit to ear.

Like all life, life in the sea depends upon solar radiation for its source of energy. Through photosynthesis, marine plants, of which by far the most important are the minute drifting organisms collectively know as phytoplankton, store this energy in a form which other organisms such as zooplankton and the fish which prey upon it can utilise, a simplified diagram showing the board outling of what is in reality a highly complex food web.

It is helpful to realise that at each step the efficiency of conversion of food into body weight is only about 10%. Nelson Smith estimates that man would require at least one kilogram of phytoplankton to gain one gram in weight. This is a useful statistic for it highlights the fact that direct effects on fish stocks eaten by man are of much greater economic importance than equivalent damage in terms of biomass lost, to the planktonic organisms on which the fish depend.

The food web diagram also highlights the fact that fresh, and other organisms well removed from the primary production stage represented by the phytoplankton, have by proxity, consumed relatively enormous quantities of organic materials some of which, although only present in small concentrations in phytoplankton, may collect at the higher tropic levels of the food web. The question of whether there is an accumulation of petroleum hydrocarbons compounds, if they represent a credible threat to the resource or the consumer, and the other possible effects on marine life of oil and of chemicals and methods used to clean it up.

Plankton

At almost all major oil spills, lay men and ecologist alike express fears about possible harmful effects on plankton. Concern can usually be summarised under three headings:

— A layers of oil on the surface will, by occluding light, cause a serious drop in photosynthesis by phytoplankton.

— The oil layer will interfere with gase out interchange at the sea surface and dissolved oxygen levels will thereby be lowered.

— Oil and dispersants will exert serious toxic effects on plankton.

How real are these effects and what is their ecological significance? The first and second are the indirect consequences of measurable physico/chemical phenomena and will be treated first.

1. Occulsion of Light

Ottway studied the light transmitting properties of 20 crude oils made up as 5% solutions in white spirit and showed that different crude oils vary widely in this respect. Oils high in asphaltenes and sulphur were found to absorb most light and, for all oils, the absorption spectra exhibited maxima at the blue end of the spectrum (325—350 nm), indicating that the 44 same group of compounds is mainly responsible in all of the oils tested.

A number of photosynthetic pigments found in phytoplankton also have absorption maxima at the blue end of the spectrum. Potentially, at least, their photosynthetic activity could be impeded by crude oils spilt at sea. It is important to remember the scale of the problem. A huge slick of black oil covering 2000 square miles of the North Sea actually covers less than 1% of its area. It is safe to conclude that occlusion of light by spilled oil, a phenomenon easily demonstrated in the laboratory, has no ecological significance for the plankton of the open area.

2. Effects on Dissovled Oxygen Levels

The solubility of oxygen in most hydrocarbon mixtures is similar to that in sea water. An oil slick is not, therefore, an impenetrable barrier to gaseous interchange. For a full discussion of this issue, the reader should consult Nelson Smith; suffice it to say that in a 2 day trial, Brown and Reid found that boiled water, beneath a 17 mm layer of oil, absorbed 73% as much oxygen as the control.

Confirmation that oil slicks do not interfere significantly with gaseous interchange at the surface of the open sea is provided, by an observation by the Scrips Institution of Oceanography, on dissolved oxygen levels below a heavy oil slick in the Santa Barbara Channel. The oxygen saturation beneath the slick was 98.5% of that in water outside the slick area.

3. Toxic Effects

There is abundance evidence that freshly-spilled crude oils contain low boiling point substances which are actually toxic to the planktonic organisms. The most toxic of these are water soluble aromatic derivatives. Low molecular weight alkanes (10 carbon atoms or less) may induce narcosis in high concentrations but these are unlikely to result from a crude oil spill in the open sea.

The aromatic solvents used in early emulsifiers were also found to be highly toxic to planktonic organisms at levels of less than one part per million. In latter formulations of equivalent dispersing power, the use of non-aromatic solvents has reduced toxicity by over two orders of magnitude.

Useful as laboratory assessments of toxicity can be for comparing the acute toxicity of oils and dispersant chemicals to small number of organisms they are rarely a reliable guide to effects on wild populations. In that such tests are necessarily short term they under estimate toxicity, but on the other hand, the toxicant concentrations used in laboratory experiments are rarely maintained for any length of time in the sea.

Evoporate losses and the rapid dilution of residual, soluble compounds ensures that toxic effects are short-lasting, even for high toxicity, solvent emulsion of the aromatic type. Thus although highly localised short-term effects on plankton were observed following the massive use of high concentrations to disperse Torrey canyon oil. There was no evidence that any planktonic species was significantly affected the population level.

The Torrey canyon experience was quite exceptional in the amount and toxicity of the materials used to disperse the oil, and even there it would seem that the rapid evoporation and dilution of the toxicity combined with the disperssed nature and rapid trunover of the plankton to minimise the ecological impact. For other major spillages, including the notorious Santa Barbara blow-out the toxic effects on the planktonic tissue have proved to be virtually detectable. The plankton of milfold Haven was examined to see if the development of a major oil port had any effect.

The report stated that changes which have occured are no greater than might have been produced by natural environment changes and population dynamics. Sanbourn suggests however that the effect on benthic larvae when they are in upper layer of the ocean may be more longer lasting as benethic animals require years to reach maturity and so the effect on the larvae may endour longer.

The only planktonic species which may conceively be in any ecologically significant danger from the toxicants associated with crude oil or dispersants are those which form fairly localised aggregations. Arrow worms fish eggs, and larvae may come in to this category. The likelyhood of ecologically significant effects in such cases is examined in detail.

Finally, to keep our assessment of toxic effect on plankton in their true perspective it is helpful to consider the scale of plankton production of in the sea relative to the that of petroleum spilled. Fraser estimates annual world phytoplankton production to be about 150,000 million tons and that of zooplankton 15,000 million tons. By contrast estimates of annual losses of petroleum in the sea vary from 2.5 million to an improable maximum of 10 million tons. Only a fraction of this petroleum is toxic to plankton and very little of this is ever likely to come into contact with planktonic organisms.

Fisheries

As was said earlier, the purpose of this chapter is to arrive at an assessment of the effects of oil pollution. An environmental stress, for which man is largely responsible, on the biology of the sea. It is important, in this context, to consider the particular case of sea fisheries, a sector of the marine environment in which man already exerts a direct and measurable influence. The term, fishereis, has a variety of meanings, the widest embracing all exploited stocks of marine organisms from macro-algae to whales, and the narrowest exploited species of fish alone. In this chapter, the term is taken to mean the fish and shellfish of commerce.

Oil pollution is potential treat to these resources for three main reasons:

— Pollution may directly harm fish and shellfish stocks.

— Pollutants adhering to, or accumulated by, fish and shellfish may render them unaceptable to the consumer.

— The presence of spilled oil may interfere physically with fishing operation.

Controlling Pollution

Now that the nature of pollution and some of the sources of this undesirable product of society have been considered, let us concern ourselves with some of the avenues open to us to accomplish some sort of control. The most obvious control strategy that can be used is to some how or other completely prevent the introduction of the pollutant into the environment. This can be done by not producing the pollutant or by restructuring the process so that the pollutant is no longer a by-product. Let us consider a few major sources of pollutants and some possible methods of control in the particular cases.

The most obvious solution to the problem of maintenance dredging is to cut off the sediment supply to the areas where sedimentation is a major problem. This can be done in a number of different ways. The most effective of all is to prevent upstream erosion, but at the same time this is perhaps the most difficult. The entire watershed must be policed very carefully to make sure that the natural landscape is change to such an extent that erosion no longer occurs.

This means extensive terracing and replanting of areas that have been denuded for on reason or another. Of course, this also means that extreme measures must be taken in areas that have no natural cover, such as agricultural areas or areas where construction is underway. A certain amount of this is being done and these are a number of legal requirements for persons engaged in construction or agricultural activities.

Another method of preventing sediment from reaching the undersired area is to provide settling ponds. If a stream is identified that carries a large sediment load, a pond and dam arrangement can be built so that the velocity of the stream is decreased to zero and the suspended sediment will have a chance to settle. The water is released after it has dropped the major portion of its load and is allowed to proceed downstream.

The solution since the ponds will fill up rather rapidly, requiring either building of new pounds or dredging of some sort. Sometimes the sedimentaion that occurs in channel area is a result of sediment loads that are introduced into the marine environment from nearby rather than those carried from far upstream. This occurs, for example, when a chennel is dredged with unstable sides so that the material just tends to slough into the deeper regions. This requires continual dredging as the removed sediment is almost immediately replaced by bottom material from the sides. One method of correcting this situation is by stablization of the bottom. Adding larger, more dense material such as gravel to the bottom is a technique utilized occasinally, although this is usually more expensive than dredging so that it is not done too often. Bottom stabilization, however, is a technique that is being used more and more to prevent the movement of bottom materials within relatively small contiguous areas.

Oil Pollution

Another problem that, to a certain extent, appears amenable to prevention at the source is that involving oil. Unfortunately, the sources of oil in the marine environment are extremely difficult to delicate. Many figures have been published and, in almost all cases, there is a significant disagreement as to the amount of oil discharge into the marine environment and the sources of the oil. There is no doubt, though, that a large portion of this amount originates on land as waste product.

As would be expected, the major source of this is associated with the automobile. Some of this lost oil is dropped from leaks in cars and trucks, some is deliberately dumped as used crankcase oil, while some results from the deposition of unburned exhaust gases. From roads and streets this material is washed into the marine environment, usually without benefit of treatment.

Even when passed through treatment plants, however, petroleum rich effluents still provide problems as they are difficult to treat and they tend to make treatment for other pollutants less effective. All the experts agree that the greatest terrestial source of petroleum products in the ocean is used crankcase oil. Some go so far as to indicate that this source supplies more petroleum to the sea than any other source.

This may be true because within the last decade or so the habits of American motorist have change markedly to the extent that a large portion of them now change their own oil, discarding the used oil in a manner that probably causes it to end up in storm sewers.

The major reason for this probably is the almost complete disappearance of the used lubricating oil re-refining industry in recent years since it was not economically feasible to refined oil. However, with the modern continual increase in the price of new oil, the re-refining industry is having a rebirth so that a large portion of the oil previously thrown away will now be reused.

Another entirely different source of oil in the marine environment is that resulting from accidents that occur during shipping and transferring processes generally speaking. The more handling steps involved, the more accidents and associated losses can be expected.

The figures as to how much is actually lost in handling accidents compared to that lost in normal ship procedures, such as pumping bilges, is not well known, but there is no doubt that a very large portion of the oil reaching the marine environment is the result of poor handling procedures in both the operation of vessel and the transfer of oil from one carrier to another. Better procedures are need, but perhaps even more than better procedures, structure enforcement of accepted safe procedures would go a long way to decreasing the amount of oil that enters the ocean each year.

When a tanker empties its load of oil, the dynamic characteristics of this vessel are markedly changed. Its weight is decreased about a thousand times and therefore it becomes extremely difficult to derive the ship, since it was designed to be driven with a full load.

Therefore when most tankers do not have a load of oil, they will fill their tanks with water for ballast. When they return for a fresh carg of oil, they must pump this water from the tanks, releasing large amounts of oil as the residue left in the tanks is washed out. There have been many schemes suggested for decreasing the amount of oil that is wasted in the ballasting-unballasting process; however, it is not clear at this time how effective these processes are or how effectively any requirements can be enforced.

Modern supertankers are extremely large vessels and therefore have major structural weaknesses, so that occasionally they will break up in unusually heavy seas. They are simply not designed to be utilized under unusual storm conditions, so that there will continue to be losses of this type in rough weather. In passing, it should be mentioned that once oil is spilled at sea under large wave conditions there is no way for this oil to be retrieved.

On the other hand, oil may be retrieved inharbour areas where the sea surface is relatively calm, but only when there are waves no higher than approximately one meter. The only obvious way to prevent oil from getting to the marine environment when a ship founders is to prevent the accident from occurring in the first place. Alternatively, once the accident has occurred the oil must be prevented from escaping from the ship. It is somewhat doubtful whether these solutions will ever be possible.

Recycling and Reclamation

If it is impossible for one reasons or another to prevent the introduction of a pollutant into the system, it may be possible at some place along the line to extract this pollutant and rescue it. It should be kept in mind that although "consumer" is a commonly used word, there is no such thing as a consumer. The word tends to imply that material is being used up, while, in actuality, it is just borrowed for a while and returned. It may be in a somewhat different form or perhaps a somewhat different shape, but nevertheless it is still returned. We don't destroy any of this mass; we just change its form, so that the general concept of recycling or reclamation is extremely appealing, especially when we consider our larger solid forms of waste. If we could find some way to utilize all of these materials, we would solve a number of our problems especially those involved with space.

Sometimes, though, it is not necessarily desirable to solve a space problem sometimes the space problem works to our advantage. For example, it has been found that biological communities living in the sea require a certain amount of physical shelter. Some species just like to be alone, whereas in other cases the young of the species require places to hide to protect against predation.

A marsh or wetlands area is a good breeding place because both of these conditions prevail. It has been found that in many cases these conditions can be created artificially by introducing large juck articles such as old automobiles or worn out ships to form effective artificial fishing reefs.

There are disadvantages to using waste products for this sort of activity. One is that in many cases these junk objects are trasnported to undesirable areas by dynamic oceanic forces. Pollution is not controlled when automobile fenders and doors wash up on swimming beaches.

When creating a fishing reef, steps must be taken to retain the integrity of these reefs for a long period of time. If proper attention is paid to the magnitude and nature of oceanic forces, artificial reefs can be made relatively permanent.

Inactuality the whole recycling and reclamation business is predicated on cost. If it is economicaly feasible to recycle or reclaim material, it will be recycled or reclaimed. If it costs more money to use the recycled material than new material, then the material will not be recycled.

It is difficult to imagine that certain materials, such as glass, will ever be recycled to any great extent because silicon is one of the most common elements in the earth's crust and it costs just about as much to make a new bottle as it does to recycle an old one. However, in the case of aluminium there is about a factor of five involved in recycling coast as opposed to manufacturing from scratch.

With the present state of the economy where things become more expensive from year to year, it appears that more items will find themselves on the list of materials to be recycled or reclaimed. Industrial processes will change as it becomes more feasible to extra materials for the waste stream from a cost standpoint and, at the same time, industrial waste streams will become less and less polluted.

Even in the area of multiple waste product use, such as the combination of water heat from either manufacturing or electrical power generation and waste nutrients from sewage treatment plants for the purposes of aquaculture, cost will always be the controlling factor.

It has been known for some time that evaluated temperatures and the artificial addition of nutrients will increase growth rates of many organisms markedly. The choice of which organism to grow is involved not only with which would be best to grow, but also which there is a market for. Setting up an aquaculture system involves picking organisms which respond favourably to the environment, developing a system by which the optimum growth rate can be produced, developing a marketing system or that the end product will be sold at the maximum possible price, and assuring the continuing existence of market for everything produced.

Not only is science included but also sociology and economics must be considered, and it is probably in these latter two areas that the major difficulties reside. Consequently, no sophisticated aquaculture systems using both waste heat and nutrients have been developed at this point, but it is probably only a matter of time before we will have extensive aquaculture using man's waste products.

Storage of Pollutants

Another method of controlling pollution, utilized for certain pollutants, is to store them at some distance from man's activities so that they are as inacessible as possible to the major fraction of the population and pathways to man. Thus we essentially throw them away, in a place where no one can get to them and their effects cannot get to man.

Occasionally nuclear waste products have been dumped in deep oceanic areas in the hope that when they do break out of their canisters, they will not find an easy pathway back to man since the deeper life forms are probably not in any food chain involving man. Perhaps not so well known is the fact that deep ocean disposal has been used for other materials such as coverage ordinance and poison gases for which the military has not further use.

In the case of nuclear products, there is a very vociferous school of thought that believes storage facilities should allow waste retrieval if a technological break through is ever accomplished that would allow these nuclear waste products to be profitably utilized. Another school feels that they should be placed in an areas as completely inaccessible as possible to everyone. This latter school in tuitively seems the safest, but, on the other hand, most common storage places are just not the type. There is always a small probability of a pathway back in man being found.

One suggestion for a nonretrivable storage place which holds some promise has been made by Issac Asimov. According to modern Geological theory the earth's crust is composed of a number of separate plates, all moving with respect to one another. In some places the plates are moving together and in others apart, following the motion of the mantle, the layer of material underneath the earth's crust. As a result of convention, the mantle tends to rise toward the earth's crust in certain areas, producing oceanic ridges such as the North Atlantic ridge. At these divergence areas, where the mantle material is brought up to the earth's crust, the motion becomes horizontal and the crustal plates are carried away from the ridges at a rate of a few centimeters per year.

The convergent regions where the plates come together are where the mantle material is carried back down toward the centre of the earth. In direct contrast to the regions of divergence associated with ridges, the convergent areas are associated with oceanic trenches where crustal material is carried down and under the continental plates. These oceanic trench areas are very deep, often as deep as 10,000 meters, and at the same time are actively engaged in transporting material into the bowels of the earth.

Controlling Pollution by Zone

Another method of controlling pollution is by specifying certain areas that can be utilized for effluent discharge and other areas that cannot. This method has been used on land as records have been kept. Although zoning of certain regions as industrial, commercial, or residential is a relatively recent innovation, there always was city dump. Now-a-days we are even starting to convert our city dumps into usable areas, such as gulf courses or sky slopes by convering, planting, and landscaping as rapidly as possible. This is an exciting concept that is gaining in popularity as dumping areas become more and more scarce, but it will be some time before ocean dumping areas are reclaimed in the fashion.

Even though we are attempting to utilize our dumping areas for other purposes as rapidly as we can, the concept of zoning still pervades our society. Certain areas are set aside for certain types of use. Society is willing to accept a factory in a particular location where it is hidden, while at the same time people will not accept that factory as a next door neighbour. The same concept has been suggested for marine coastal areas.

Certain regions would be reserved for industry, others for sewage treatment, while still others would be set aside for recreation. Some of these activities are such that they may be pursued in conjunction with one another, whereas others cannot. The idea is for every coastal areas to try to obtain utilization of marine resources available.

In order to accomplish this type of zoning most effectively, a great deal more knowledge, is required than is presently available about the synergistic effects of various types of activities. One of the major advantage of optimum zoning is that it would allow maximum avoidance of undesirable synergistic effects.

The biggest stumbling block to coastal zoning probably is in overcoming entrenched interests in regions where the optimum activities are different from those presently underway: A socio-economic problem rather than a technological one, the Coastal Zone Management Act encourages coastal states to attempt to take the first step in planning for optimum utilization of the coastal areas.

After plans are complete, the states are to implement them with the financial assistance of the federal Government. The concept is an interesting one, because it essentially allows some pollution of certain areas while refusing to permit any pollution of other locales.

This is somewhat different from the way pollution is presently controlled, where the effort is made to limit pollution in all areas to the minimum possible amount. With the zoning concept pollution would still be limited as much as possible, but it would be permissible to allow the water quality to degrade below the mean in some places, since in others it would be higher than the mean.

Controlling Pollution by Taxation

One last method of pollution control that has been suggested, although it has met with a great deal of resistance and antagonism, is taxation. As is well known, taxation can be utilized to encourage or discourage certain activities in addition to raising funds to support government programs.

Pollution would be allowed, but it would be taxed at a rate proportional to the environmental insult. This general philosophy is followed at the present time by requiring the polluters pay fines roughly proportional to the amount of pollution, and in many cases industries will purposely pollute and pay the fine because it is apparently cheaper than making the necessary plant modification required for cleanup.

Obviously the system is not working if this is the case; the fines are too low. Some economists suggest that a tax on pollution will simply be a license to pollute and will therefore encourage pollution rather than discourage it. Other economists, however, indicate that pollution taxes should be based on the severity of the pollution with some sort of graduated tax.

A little pollution requires a little tax, a large pollution requires a large tax. An important aspect of this concept is that the words the true and total cost to society of these pollutants. Every time the common resource is utilized by an individual, this society for the right to use this common resource.

This payment would be in the form of a tax set at a level determined by how much society economists believe that anyone utilizing the resource should pay for its degraded then the individual degrading should pay for the degradation. This is an interesting concept that is being discussed in economic circles taxation certainly is a powerful tool for societal behaviour modification that has worked many times.

Generally speaking, taxation encourages particular activities and discourages other if the taxation amount is large enough to completely discourage undesirable activities.

From a management point of view there are a number of different directions that can be taken to control pollution, including both technological and socio-economic alternatives. So far society to aim more in the direction of the technological alternatives, but there appears to be definite swing toward other methods. In the year to come it appears that more and more emphasis will be placed on these non-technical methods especially as the trade offs, costs, and benefits become better understood.

The Forces Involved in Dispersing Pollutants

Once a pollutant has been introduced into the marine environment, it becomes a matter of no small interest to determine where this pollutant goes and how the concentration of the pollutant varies with time and location. These process fall into two basic groups: advection and dilution, advection is the mechanism by which the pollutant is carried from one place to another, while dilution results when the pollutant is caused to reduce its concentration by some sort of spreading mechanism. Advection simply has to do with oceanic currents and will include three primary types of current since these are encountered most often in the oceanic regime.

Management of the Marine Environment

Up to this time we have been primarily concerned with the technical aspects of pollution. Space has been devoted to discussion of what pollution is, where it comes from, and where it goes in the marine environment. Now we must consider how to control this pollution so that the problems resulting from it are kept to a minimum. The previous sentence has a number of words in it which should probably be defined very carefully, but probably never will be.

For example, the statement that pollution should be controlled so that the resulting problems be kept to a minimum may be interepreted in many different people since a problem for one group is not necessarily a problem for clear-cut.Just as one man's trash is another man's treasure, one man's pollution is another man's enrichment, and this is not the only dichotomy that faces a conscientious marine manager. There are problems of jurisdiction involved in whether the responsibility for running the operation lies with the federal government, a local government, or some regulatory commission, and always there is the application of priorities supposedly set by society as a whole. The basic tool the manager has to work with is the law, however, and this is where we begin.

The Legal background

The legal background for pollution control goes back to the Rivers and Harbuors Act of 1899, recently interpreted so as to be used as a pollution control measure. This act set up responsibilities for protecting navigational and environmental aspects of all navigable waters to the U.S. Army Corps of Engineers. The Corps therefore has the responsibility not only of protecting navigable waters so they remain navigable but also protecting them against their loss as a viable vehicle for various other uses that society considers desirable.

An interesting aspect of this law and one reason it has become such a powerful tool in the hands of the environmental protectors is the redefinition of navigable waters. It turns out that the modern definition includes just about every stream going into a large river in the United States. Thus the Army Corps of Engineers was given the responsibility for protecting the environment 1899 and to all intents and purposes still retains this primary responsibility as far as brackish and salt water is concerned.

In 1934 the Fish and Wildlife Coordination Act was passed. This has since been amended many times, but essentially it involves the fact the U.S. Fish and Wildlife Service and The National Marine Fishers Services have the responsibility for determining the effect of any proposed project on fish and wildlife resources. They then report to the crops of Engineers which makes the final decision as to whether or not the proposed project is allowed.

In 1948 the Federal Water Pollution Control Act was passed and this also has been emended periodically. It requires the Corps of Engineers to apply environmental standards and criteria in approving the disposal of spoil in navigable waters. The most recent amendment of this act was the Federal Pollution Water Control Act of 1972. This a complete rewrite of the 1948 law and it set out four basic requirements. In the first place, a national goal for the elimination of all pollution from America's waters by 1985 was proclaimed. This is the famous zero discharge goal.

Note that this is not a mandate to have zero discharge by 1985, but as a national goal it is something to aim at and it is to be hoped that we will come as close as possible. Second, the Act of 1972 set as another goal the addition of secondary treatment to all municipal sewage treatment plants by 1977. This goal has not been met but there has been a strong effort in that direction. Third was the effort to encourage the use of new, more advanced treatment and disposal methods by 1983 so that some of the problems will be resolved by the use of advanced technology rather than by simply improving on present methods.

And last, the act established an industrial cleanup program with restrictions becoming tighter with the passage of time. Each year the requirements involving effluent standards will be tightened so that these will be continuing cleanup of the nation's waters. This particular aspect of the bill was backed by stiff penalities including both fines and imprisonments.

This 1972 Act is the basic law under which most antipollution suits are brought. The Water Pollution Control Act of 1948 encouraged the creation of uniform laws to control pollution, supported research, or firms to cease practices leading to pollution, and established the

Federal Water Pollution Control Advisrory Board. This has been succeeded in recent years by the Council Environmental Quality.

The year 1965 saw the passage of the Water Quality Act. In this act states were given an opportunity to adopt and enforce federal water quality standards for interstate waters. The Clean Water Restoration Act of 1966 authorized massive federal participation in the construction of sewage treatment plants. About 3.5 billion dollars were allocated to be spent between 1967 and 1971.

The National Environmental Policy Act of 1969 provided preparation of environmental impact statements when the U.S. Army Corps of Engineers determined it to be necessary. The Act also declared it to be national policy to encourage the production of enjoyable harmony between man and his environment.

The Water Quality Act of 1970 addressed oil pollution from vessels and offshore facilities, federal permits, sewage pollution from vessels and hazardous substance discharged into the nation's waters.

Also in 1970 recognization plan No. 3 was submitted to the Congress by the Chief Executive. This reorgnization plan established the Environmental Protection Agency, setting up the organization and transferring responsibility for various activities from other agencies. Unfortunately, the executive order does not address the problems of the objective or the purpose of the Environmental Protection Agency (EPA) and this objective or purpose has been left to the Agency Administrator to interpret for himself. This omission has resulted in a general grey area with EPA being as strong or as weak as the Administrator and Congress want it to be. In 1972, in addition to the amendment to the Water Pollution Control Act, two other act of some import to the protection of the marine environment were passed.

The first of these was the ocean Dumping Act of 1972. This act requires the Army Corps of Engineers to apply Environmental Protection Agency standards and criteria in approving ocean disposal of dredge spoil materials where transport passes through United States territorial waters. This is an interesting law because it controls the material that originates in United States ports and not that material which is dumped in United States water. Consequently no matter where the material is dumped, as long as it originates in a United States port, the Corps of Engineers has some control over it.

The other act passed in 1972 having important ramifications to the marine community was the Coastal Zone Management Act. This act gave the individual states primary responsibility for determining land and water uses to more effectively balance environmental protection and economic development objectives in the coastal zone. It delineated the coastal zone as including all navigable waters and defined them as all waters up to the mean high tide line, including lands wholly or partially covered at high tide, whether privately or publicly owned.

Under this act large amounts of money have been given to participating states, and all coastal and Great Lake states have so far indicated a desire to participate. Funds are initially for planning purposes aimed at developing a general zoning plan for the use of the coastal zone. Later funds will be available for the implementation of these plans.

In an attempt to get at specific causes of pollution, the Ports and Waterways Safety Act was passed in 1972 which addressed the problem of pollution from ships. This act allowed for ship design and construction regulation with respect to possible environmental damage and it applied to all United States flag vessel as well as all foreign vessel entering United States waters carrying either oil or hazardous liquid cargo in bulk. Vessel design and operational standards for all these vessel in United States waters were finally formulated on 23 January 1977. In this manner the United States government is attempting to prevent as many large catastrophies involving vessel carrying oil or hazardous materials as it possibly can by simply not allowing marginal vessel to navigate within the territorial limits of the United States.

Another fact of marine pollution having a source outside the continental area is involved with driving for oil and gas offshore. The Outer Continental Shelf Lands Act gave the responsibility of Granting leases for continental shelf drilling to the Department Interior. These leases require the leases to reimburse the federal government for any oil spill cleanup cost and in addition they require the installation of blowout prevent equipment on the wells themselves.

International Convenants

The foregoing is a brief list of the basic framework of federal laws which govern the discharge of waste products into our terristorial waters. In addition there are a few international convenants under which the United States operates.

The Intergovernmental Maritime Consultative Organization (IMCO) has sponsored three conventions resulting in a number of regulations. The first of these was the 1954 Oil pollution convention which was amended in 1962 and 1969. This convention places limitations on the rate of discharge, the oil content, and the distance from land for oily ballast discharge water from ships. The enforcement of this convention is left up to the flat state, that is the country where the ship is registered.

The second convention was the 1971 amendment to the oil pollution Convention which branched out in different direction by addressing vessel design. Unfortunately this part of this amendment has not been rectified by any maritime nation.

In 1973 a ship pollution convention was sponsored by IMCO, and this convention resulted in a set of regulations establishing standards for ship board sewage treatment equipment and also specified the requirement for segregate ballast tanks for new ships of specified sizes. It also specified methods for storing dirty oil ballast residue, oil discharge monitoring systems, and oily water oil extraction systems. Unfortunately this convention also has not been ratified by any martime nation. Consequently, behaviour on the high seas is left strictly to the conscience of the individual ship's master.

At this time there are also no international agreements on ocean dumping or the environmental effects of seabed resource exploitation. There have been many attempts to draw up an International Law of the Sea agreement but as of this writing none has been successful.

Jurisdictional Responsibility

One of the problems facing a marine resource manger is the problem of jurisdictional dispute. Who does have responsibility? In the four general areas the responsibility is somewhat dispersed. The first of these areas is water drainage, capture, or use which is exclusively on a local level. Since this is primarily a problem of private rights which is usually involved with state of private ownership. This area does sometimes get involved as a pollution source so it is of concern to us. The third area of interest includes the bays and coastal waters which are controlled by a combination of state and federal regulations. Generally speaking, the state has control over resources and the federal government has control over navigational and environmental considerations, but there are many exceptions. Most state laws are not quite as stringent as the federal environmental requirements so that the federal requirements must be obeyed. However, there are some state laws more strict than the federal and when this is the case, the state laws take precedence. The last areas is the region beyond the coastal region wherein the responsibility is somewhat loose.

We have seen that there are no viable international agreements for the deep ocean, but in many cases the United States has acted unilaterally to protect its continuous waters. With the passage of the Ports and Waterways Safety Act and the Outer Continental Self Lands Act attempts have been made to protect against the effects of oil and hazardous material, cargo ships and offshore oil-wells.

Political Processes

Once the problem of jurisdiction is settled, there then appears the problem of political processes. The political process is such that technical standards will almost always yield to such things as hardship cases, emergency situations, or strong public sentiments. Consequently, any decision to be made where the environment is involved must have a public input or else the decision will probably not be effective.

The manager must not only face an often emotional public, but also individual agencies, both state and federal, working at cross purposes. For example, the Environmental Protection Agency is primarily concerned with water quality, the U.S. Fish and Wildlife Service is primarily concerned with fish habitat improvement, while the Army Corps of Engineers has a primary concern in the areas of navigation improvement. These concerns must be balanced against common public needs such as high employment, protection of science environments, and housing demands, very often these conflict very strongly so that at least one of the many groups must eventually compromise. This compromise, in the long run, must be based on some sort of a set of societal objectives.

Society must somehow or other determine the most important things it wants from its marine resources. The activities range from aesthetic enjoyment of the area to commercial shipping and waste disposal. Obviously waste disposal is not exactly the type of activity to enhance aesthetic enjoyment, nor is utilizing the coastal zone for military maneuvers encouraging to the maintenance of a reguge or a sanctuary. Some of the uses of the marien coastal area must be sacrificed in certain cases for other uses which society deems more important.

Somehow or other society must set priorities so, that the areas can be managed to obtain the ends that society wants from them. This is a very important point and one that is often completely ignored. In very few cases has society overtly set down priorities for the use of marine areas, and until priorities are specifically stated, marine area managers will be forced into making bad decisions. It is to be hoped that some of the planning being done under the Coastal Zone Management Act will involve priority setting, especially regarding the use of particular areas. If this is the only outcome of the coastal zone management act, it will indeed be a valuable contribution to the management of the activities of society in the marine coastal zone.

Costs, Risks, and Benefits

In order to set priorities in a rational manner, managers must by one means or another determine benefits, costs, and risks associated with any and all activities. Once these three parameters have been quantified, then cost7 benefited rations may be determined and priorities may be set on the basis of these ratios. However, this is much easier said than done. All three of these parameters are usually quite qualitative and very difficult to quantity. Benefits, for example, are often what people perceive them to be and therefore are essentially a political evaluation rather than an exact calculation.

Costs can often be determined accurately in terms of the actual physical nature of the activity. Buildings, machinery, and labour can be quantified in terms of their cost. However, very often an attempt to control pollution or the effect of pollution on society results in costs that simply cannot easily be quantified.

Risks also are difficult to evaluate. How much is a life worth? Does it make a difference if it is yours or someone else? Thus we find that we must be satisfied with some approximation to these three parameteres. Some methods for obtaining approximations to values for costs, risks, and benefits will be discussed below.

The cost versus pollution control curve, is an exponential curve wherein the initial costs of controlling pollution are relatively inexpensive, but the final 5 or 10% of pollution control may cost as much as the initial 90 to 95%. The benefit pollution control curve, on the other hand, is parabolic in shape. Here the benefits in the initial cleanup phase are quite large while the benefits acccured from the final stages are quite small. We might consider then some sort of a cost/benefit ratio, knowing these two functions.

The optimum cost benefit ration is determined on the basis of the incremental cost per incremental benefit. If, example, an additional dollar of environmental control will produce more than a dollar's worth of benefits, then it is more effective to spend additional money on control. If, however, additional control results in fewer dollars worth of benefit, then it appears that the control process has been carried too far.

Table 2

Risk Table

Risk	*Death rate per million hours of exposure*
Voluntary	40.0
Rock climbing	6.6
Motorcycle	2.4
Scheduled airline	2.4
Smoking cigarettes	1.2
Death by disease	—
Including old age	1.0
Private auto	0.95
Sking	0.71
Railroad and bus	0.08
Involuntary	—
Electric power	0.002

The optimal level of pollution control is not at point A where the cost of pollution control and benefits are the same, but is at point B, where the rate of increase of cost per unit pollution control is equal to the rate of increase of benefit per unit pollution control. From a cost benefit analysis then it appears that it is not desirable to completely clean up the environment since the benefits accured would not be worth the expenditure required. However, this is not the complete story. It is obvious that in certain cases even a small amount of pollutant is not acceptable because of the danger to health. Thus the risk too much be considered.

Each activity and each pollutant has an associated risk, some of these to the environment and some to man. It is difficult to quantify these but one method commonly used is to determine the number of fatalities per million hours of exposure for each person involved in the activity considered. This is expressed as a statistical probability. Table 1 is a listing of risks expressed in this manner for various common activities.

Notice that the activities are divided into two groups, voluntary and involuntary activities. The voluntary activities are those which the individual can decide for himself whether or not he wishes to do. Notice that many of these activities have a probability very close to that of death by disease, including old age. Others, however, such as rock climbing, have an extremely high risk associated with them and therefore are only for the adventuresome.

Interestingly enough, the risk associated with the one involuntary activity listed, electric power, is less by about three orders of magnitude than most of the risks in the voluntary portion of the table. Apparently then we are loath to let others do in to us what we happily do to ourselves.

Pollution, of course, falls within this area of involuntary activities since any pollutant discharged to the marine environment is not directly controllable by any individual. Consequently we would expect that the risks associated with pollutants would be down in the area of that of electric power rather than up with that of railroad and bus travel.

In order to quantify priorities, it if necessary then to develop some sort of risk benefit ratio, which, in turn, requires quantification of benefits. One possibility of quantifying benefits is to simply assume that in the case of voluntary activities the amount of money spent on the activity by the average involved individual was proportional to its benefit, while in the case of involuntary activities, the contribution of the activity to the individual's annual income as porportional to its benefit. Here risk is plotted relative to the benefit for various kinds of voluntary and involuntary exposure.

From these graphs and some other works that has been done, a number of conclusions can be drawn. First, the public is apparently willing to accept voluntary risks roughly a thousand times greater than those involved with involuntary exposure. Second, the statistical risk of death from disease and old age appears to be a psychological yardstick for establishing the level of acceptability of other risks.

Third, the acceptability of risks appear to be crudely proportional to the third power of the benefits, whether they be real or imagined. Fourth, the social acceptance of risks appears to be directly influenced by public awareness of the benefits of an activity as determined perhaps by advertising or the number of enhanced by increased public awareness.

A determination of the level of optimum pollution control and general management schemes involved with addressing the total pollution problem is not as easy one. We find that everything costs something. There is no such thing as a free lunch. For every cleanup process involved there is a cost. Sometimes this cost is minimal but other times the cost is extremely high. For example, it has been estimated that in order to meet the 1983 goals of the Federal Water Pollution Act or 1972 it would cost 350 billion dollars just to trust municipal sewage in an adequate manner. This includes the cost of controlling nonpoint sources along with sewage treatment plant updating.

Whether or not the America public is willing to spend this much on this sort of program is at this time unclear, but the American public must be media ware of the necessity of costs and trades of any cleanup program.

If we are going to spend this amount of money for pollution cleanup then there will not be as much money available for other activities, and once again we must decide which to these activities, and once again we must decide which to these activities is more important to us. But the decision can only be made if we have as much data as possible in hand to allow as accurate as possible a determination of benefits, costs, and risk involved with each process.

The Future of the Marine Environment

In the previous pages we have seen that the oceanic volume is extremely large, while at the same time, that portion of the oceanic volume available for pollutant discharge is relatively small.

Consequently, we hope to have demonstrated a need for some kind of marine pollution management since there is a very real problem associated with dumping wastes in the ocean. It has also been shown that an ever increasing population, distributed primarily around coastal areas, will inevitably increase the total amount of waste discharged to the oceans.

This is coupled with the fact that technology is continually growing and its growth will be associated with an increasing amount of pollution per capita with the passage of time. It all boils down to the realization that the total volume of water available for waste dilution will remain relatively constant while the total amount of waste will probably increase as far into the future as can be foreseen at this time.

Management Requirements

Fortunately, however, in those nations which have reached a relatively high level of technology, there is an increasing awareness of the problem and a public requirement that planning and management proceed in order to utilize the marine resource in the best manner possible. In order to do this a number of requirements must be met.

In the first place the effects and pathways of all pollutants dumped into the ocean must be known. This is an extremely difficult task. Many of the pathways are known but the effects are not. For other pollutants the effects are well known but the pathways back to man are still mysterious. This obviously will require a great deal of research, and even then there is no guarantee that all effects and pathways for all pollutant materials will ever be known.

Second, the average residence time for all materials discharged into the marine environment should be known. Here again we are faced with an almost impossible problem due to the tremendous amounts and number of different materials discharged into the oceanic environment.

There should also be a continuing monitoring of the environment if only to determine gross changes so some action may be taken before the situation gets out of hand.

Lastly, we must continually determine and redetermine what people consider to be important. The more and folk ways of society constantly change, altering values and priorities.

2. Setting Priorities

At this time managers have no clear understanding as to what people's priorities are, nor is there any real indication that the people themselves have any great insight into their needs and desires.

We do not know, for example, whether the major portion of the populace considers sport fishing or ocean dumping to be higher in priority with respect to usage of the marine environment. We do know what the small special interest groups have to say about this situation, but in terms of jobs lost or consumer goods price increases which would result from the discontinuance of ocean dumping, there is very little information.

In order to determine society's needs and desires the people must first be educated as to the costs, benefits, and risks associated with any environmental protection solution being considered. Obviously a much better choice of alternatives can be made if the advantages and disadvantages may very well be a utopian dream, but it certainly should be the goal of each manager to come as close to this as possible.

At the present time choice are made on the basis of limited knowledge since not all of the costs, risks, and benefits are available to the manager making a decision, even those that are available to the manager are usually not available to the public at large so·that the manager's decision can be evaluated. This situation must change if utter choice is to be avoided.

Once our cards are on the table where they may be seen by all interested parties, we may proceed by setting goals that are both popular and attainable. A goal such as "zero discharge by 1985" is not an attainable one and when such a goal is set and not met, very often the result is a certain cynicism regarding all future goals.

3. Goals, Guidelines, and Standards

At this point it may be desirable to define some words commonly used in the management of the marine environment. In particular there are four words needing clarification: goal, guideline, criterion, and standard. Although the definitions which follow may not necessarily agree with those in a standard dictionary, they follow the usage of most people involved in marine area management, standard: A plan established by governmental authority as a program for water pollution prevention and abatement.

— *Criterion*: A scientific requirement on which a decision or judgement may be based concerning the suitability of water quality to support a designated use.

— *Guideline*: An acceptable methodology for achieving any given standard.

— *Goal* : The ultimate standard which all current standards may eventually approach.

4. Involving the Public

Thus goals must be set with two criteria in mind. One is the scientific requirement defined above, but there is also a personal requirement on the part of the public based on its sense of priority. The public will not support a particular cirterion, even with substantiating scientific data, if it interferes markedly with the present lifestyle.

Until the citizenry can be educated to the risks involved is not accepting scientifically based criteria, looser criteria must surface. But it is the duty of the manager not only to choose the acceptable criteria but also to attempt to educate so that more logical criteria may receive public support. Along with the choice of particular goals, standards, criteria, and guidelines for a particular time, it must be assumed that these are not permanent but may very well change with the passage of time. Both the environment and public attitudes must be constantly monitored to effectively meet the needs of a changing society. With a proper monitoring system, not only may the standards, criteria, and guidelines be change but also the goals may be update.

In this manner the public is made a working part of the management process adding support to the process and consequently making it more effective. The Castal Zone Management Act was designed along these line and it has been in operation long enough for a major portion of coastal states to have prepared detailed plans on the optimum use of coastal areas.

These plans are undergoing extensive public hearings and it is to be hoped that when they are finally adopted the public will actively support them. If this is indeed the case, it can be hoped that the effect will, at the very least, be to stop the increase in marine pollution and may be even reverse the trend by decreasing pollution. In any event, however, regulations set down to protect the environmental quality of the marine environment will be enforced with popular support.

Impacts Associated with Cleanup Techniques

1. Cleanup Effects on Marshes

The equipment used for several cleanup techniques considered here has no impact on the marsh itself. For example, mechanical and weir skimmer are deployed in tidal channels, and pools, not in the marsh proper. Vacuum skimming in itself produce no impact on the marsh, but it is frequently accompanied by substantial foot traffic. Food and equipment traffic and considered secondary impacts and may, in themselves, produce substantial adverse effects.

2. Effects of Traffic

The most obvious type of disturbance caused by traffic in a marsh is physical breakage of plants. In both grass-dominated and succulent-dominated marshed, physical damage effectively reduces the amount of photosynthetic tissue and may expose the interior of the plants to toxic fractions of the oil. Plants are likely to recover from one such event, but recurrent trampling may clear a part that will persist for years. If the soil is soft, as it is in many marshes, roots and rhizomes may be broken and thereby accelerate erosive processes. Foot traffic potentially accelerates erosion even where the substratum is firm and the rhizome may remains essentially intact. Shorelines that consist of steep escapements are particularly vulnerable. Cleanup activities that entail foot traffic should be used in such regions only after other methodologies have been considered and rejected.

Sometimes it is necessary to transport heavy equipment to remote portions of marsh. The impacts of foot traffic described above apply even more strongly to vehicular traffic. Traffic on soft marsh soil may bury plant stems and leaves and reduce their productivity until they resprout or grow back above the soil surface. Further, traffic under these conditions is likely to bury oil. In the anaerobic soils that characterized mangroves and marshes, residual oil may persist for years. If the buried oil is toxic, it may inhibit the growth of anything in the contaminated zone until it eventually dissipates.

3. Flushing

Low pressure flushing with sea water (from a nearby source and thus likely to be of a salinity to which the marsh is accustomed) may be beneficial if used with caution. Where vegetation cover is continuous and sediments relatively stable, low pressure flushing may be effective in removing substantial amounts of oil from vegetation. Drawbacks of the method are:

— All oil is not removed from leaves and a sufficient amount may adhere to the waxy plant culticles to cause damage. This is especially true when the oil is fresh in which case it adsorbs onto leaves very strongly an penetrates them.

— Foot traffic required to deploy the flushing equipment may cause damage.

— Oil flushed to creeks, etc., will be available to re-oil the marsh or other areas unless properly collected or otherwise removed.

Where vegetation cover is incomplete where much bare mud is in evident, or where sediments are sort and muddy, sediment disturbance or erosion may produce additional damge to that of the oil. Wherever possible, sediment disturbance should be avoided. High pressure flushing may cause some erosion, local rearrangement of the substrate, and physical damage to the plants. These forms of damage may be less severe than similar impacts caused by foot traffic, but this depends upon mode of deployment. High pressure flushing may cause oil to be driven into the substrate. If steam or heated water is used, marsh animals and plants in the spray pattern may be killed or stressed by thermal stock.

Sorbents: Sorbent pads, oil snares, and similar cleanup aids have two major drawbacks:

— They are usually used by a large group of personnel who heavily traffic the marsh and

— They must be recovered. Additionally, if cleanup teams are not trained in the use of these material, there is chance that oil may be mised with a shallow layer of the substrate in the course of recovery.

Because complete recovery of sorbent materials is seldom possible in a field cleanup exercise, remanants of sorbents may persist. Most remanants are merely unsightly, although some may ensure birds and animals. Large concentrations of undergraduate materials could block the light from an appreciable portion of the marsh surface and consequently reduce marsh productivity in that region. Biodegradable sorbents avoid many of these problems. Oil sorbent materials (synthetic and biodegradable) should always be recovered from the marsh, those that are not collected are a potential source of recontamination and a hazard to marsh animals.

4. Cutting

Cutting of oiled plants is a cleanup technique that entails direct physical destruction of plant tissues. As such, it severely reduces the amount of photosynthetic tissue and may expose the interior of the plant to toxic substances in the oil. Moreover, cutting operations are likely to require a great deal of foot traffic and the attendant adverse effects. Cutting is probably most beneficial with certain species of grasses and rushes that have been heavily contaminated with viscous oil and are not subject to natural cleaning. If the entire aerial portion of the plant is coated, the roots are likely to suffocate unless some passage way is opened to the interior of the plant.

Cutting accomplishes this, provided that free oil that might replug the cut stem has been removed from the surrounding marsh. However, if oiled marsh plants are not throughly coated with oil air can still diffuse down the stem to the foots, so that cutting is unnecessary unless other threats are present. Spartina marshes are very tolerant of occasional cutting, especially late in the growing season. Saltwort marshes are less tolerant. Burning is sometimes an effective method of removing oil and contaminated vegetation from a marsh without encouraging injurious foot traffic. Spartina marshes can with stand occasional burning. In fall and winter, they die back to a state of dormancy. During this period, the plants are dry and may support burning. In fact, fall burning of marshes used for agriculture is a commonly applied management tool.

During the dieback period, burning can be achieved in spartina marshed without damaging the buried portions of the plants and can stimulate their regrowth. In all other seasons, however, not only in Spartina difficult to burn, but the growing portions, shoots, and buds are injured by burning.

However, this damage does not necessarily permanently harm the marsh since the underground portions of the buried plants are likely to sprout and replace the destroyed portions soon after the burning. Saltwort marshed do not die back seasonally and the plants do not have large, protected underground systems; thus, burning is injurious at any time.

5. Soil Removal

Removal of soil entails elimination of marsh plant habitat and should be avoided. Nonetheless, if the substrate is heavily saturated with toxic oil and no predicted to recover naturally, this may be the only available cleanup technique. On soft areas of fine mud, it may not be feasible at all.

6. Dispersant Chemicals

Dispersant chemicals have been applied to oil in salt marshes following spills and in a number of experimental situations. In many cases, these have caused additional damage to the oil alone, and in some cases can be demonstrated to enhance the penetration of oil into intertidal sediments. Recent evidence suggests that new generations of dispersant may cause little additional geological effects to oil along when used at appropriate concentrations.

However, the value of using them in most cases if unclear praticularly on extensive marshes where, following application, dispersed oil may enter creeks to affect other part of the marsh or intertidal systems. They may be of value in cleaning small and stable fringing marshed in combination with low pressure flushing, but, in general, dispersant use is not recommended for salt marshes. It should be borne in mind that spraying of near dispersants onto vegetation produces damage and correct dilution to manufacturer recommendations is vital. Marsh cleaning methods do not appear to decrease damage done by oil and often increas in. The need for cleanup should be carefully reviewed before any action is taken.

7. Cleanup Effects on Mangroves

Few publications are available that deal directly with the effects of oil spill cleanup on mangroves. The several observations of the effects cleanup on mangroves at the Florida Keys oil spill of 1975. So many authors summarize effects of cleanup of marine wetlands and include numerous general statements concerning mangroves. The following summary is based on the publications and the personal observations several scientists made during cleanup operations within oiled mangroves.

Recovery of Marshes from Oil Damage

Although a definition of recovery has been attempted, mainly to distinguish the process from that of restoration, the definition includes the terms strcuture and function. The majority of scientific literature on the effects of oil deal with changes in the structure of marshes relatively few deal with the function of marshes let alone the part oil damage may play. Hence, we may be able to determine the time at which a marsh returns to something approaching its original structure, but not necessary. Alternatively, a similar marsh function could be achieved with a different structure. This is a line of investigation that deserves further attention, but at present we can do little but recommend that marsh structure be used in the assessment of recovery.

In an earlier section, the rate of recovery of salt marshes has been shown to be considerably influenced by a large number of factors. Their precise action and interactions are often poorly known, which means that in practice each spill must be treated as unique and assessment of damage must be dealt with in each case on its own merits little more than generalization is possible from the scientific literature.

Recovery proceses begin when oil toxicity is reduced or removed. In most cases, recovery starts irrespective of the activities of man, although cleanup or other action may influence the rate. Where oil damage is relatively light and little of the toxic material remains in the system recovery can be rapid, either by regrowth from rootstock or from seeds. Baker reports that single oilings in field experiment, although toxic to mary salt marsh species, may produce effects detectable only for one to three growing seasons. Where oiling is repeated or where damage is servere and toxic materials are retained in sediments, recovery may take much longer, and the effects of the spill may be detectable over decades. Where erosion of sediments takes place, recovery may not occur.

Oil Spill Prevention and Cleanup in the Vicinity of Coral Reefs

1. Prespill Mapping

Prespill contingency planning is now an accepted part of preparation for, and response to, oil spills. Vulnerability mapping has been used by several authors to identify those area of shoreline that may be particularly sensitive to oil spill damage and to help ensure that damage due to the use of inappropriate cleanup procedurs is minimized. One frequently used vulnerability index was devised by using the assumption that coastal geomorphology frequently determines the types of ecological communities found there. Information about points of access to a spill site is essential in the event that men and equipment must be deployed at shore notice. Therefore, accessibility mapping is an important adjoint to voulnerability mapping.

Coral reef voulnerability to oil spill damage will vary with reef type, zonation patterns, and tidal actions. For example, in a fringing reef, the seward facing reef creast is generally exposed to high wave stress; here oil will probably have a short residence time before being dispersed naturally. In contrast, wave action is a lot less severe in the shallow reef flat that frequently supports seagrass beds. Here, a long oil residence time with a correspondingly high risk of oil damage is likely. Similarly, oil that strands on reefs with a limited tidal range will likely tend to persist for longer periods than that in areas with large tidal ranges.

Although vulnerability and accessibility mapping techniques are now well established, little evidence suggests that such techniques are being applied to reef ecosystems. Some reef areas are well mapped and information is readily available on the location of refineries, tanker terminals, oil platforms, and major tanker routes. e.g., International Union for Conservation of Nature and Natural Resources. This information can be used to produce maps that help locate areas of high potential risk. Newly developed techniques for mapping large areas of coastline, such as low altitude video overflights seem particularly suitable for use in reef areas but have yet to be applied.

2. Cleanup and Treatment of Spilled Oil

The sensitivity of reef ecosystems and the likelihood of their being impacted by oil will depend upon a variety of factors, including the quantity and type of oil and the degree of weathering. The likelihood of ecological impact will also vary according to whether the reef is an emergent, shallow submergent, or a deep water type. The field and laboratory studies reviews indicate that the likelihood of damage is increased if the oil is incorporated into the water column of if it comes into direct contact with the coral surface. This, therefore, places the mergent or shallow submergent section of a reef as areas of high potential sensitivity. It also indicates that any cleanup of treatment attempt should endeavour not to enhance the transport of oil into the water column or into the permanently submerged sections of a reef.

Careful consideration should be given to the treatment of oil in the vicinity of a reef as situation may develop where the cleanup or recovery attempt may be more damaging then the oil alone. Baker has pointed out that merely the shipping activity associated with the cleanup attempt may be damaging due to sediment resuspension by propellers and anchor drag breaking up corals.

Although practical experience of oil cleanup in the vicinity of coral reefs are lacking, mechanized techniques for containment and recovery of oil using booms and skimmers are acceptable as they do not cause oil to sink or become incorporated into the water column. Any restriction on use of these techniques in coral areas is likely to be brought about not by the risk of ecological damage but by other environmental considerations such as high current speed, severe wave action, or shallow water depth that make their deployment impossible.

At present, only few data are available on the impact of dispersants and dispersed oil on corals. Most information that is available has been derived from laboratory studies. Laboratory studies have also been carried out on biota associated with reefs. Both categories of study noted damage by dispersed oil, but the applicability of these in vitro data to a field situation is questionable.

Similarly, these studies used early generation dispersants which of themselves, where toxic to the corals. Using dispersants currently available, indicated that concentrations of 20 to 50 ppm dispersed oil in a 24-hour exposure produced temporary stress reaction in corals. Few deleterious effects from 24-hour exposures at 1 to 5 ppm were observed.

An oil removal technique that cannot be recommended for use in coral areas is the application of particulate sinking agents. These materials absorb oil and because they have a density higher than that of sea water, sink with the oil attached. Deleterious effects could be expected from the sunken oil coming into direct contact with the corals.

Several methods currently in the developmental stage appear to be potentially suitable for use in coral areas and other marine environments likely to be sensitive to oil. Among these are agents that have been developed that, when added to oil, gel in a semisolid form that can then be recovered. Research is also being carried out to develop nutrients that accelerate the bacterial biodegradation of oil.

Several studies have shown that corals are seemingly uneffected by a layer of oil floating on the water surface. Therefore, where unsuitable hydrographic conditions or problems of access preclude the use of mechanical containment and recovery techniques, it is probably best, given our present knowledge, to leave the oil alone to weather naturally.

Measures of Damage at the Species Level

Structural changes occur on a reef when a species or combination of species that are dominant components of a system are replaced by other species are eliminated from the system. If the replacement of elimination of a species is significant, then community, or functional level changes can become evident.

Changes in species richness, coral cover, and species diversity are potential criteria for the assessment of catastrophic structural damage to corals and any subsequent recover. Loya has modified the Shannon-Weaver diversity measure for use with corals as well as any encrusting assemblage. Grigg and Maragos used these parameters to describe recolonization by hermatypes on lava flows, as did Sheppard when assessing the impact of harbour construction on corals. Porter *et al.* used species richness, coral cover, and relative abundance to assess structural recovery of hermatypic corals following hurricane damage.

In a recent paper, pearson argued that measures of diversity, percentage cover, and similarity taken individually do not provide useful measures of either damage or recovery. He suggested that measures of damage and recovery should totally icorporate measures of percentage cover, mean colony-height, surface-index, species diversity, and similarity indices. In addition, because reefs are dynamic systems that can undergo change due to numerous environmental factors, any observed structural changes should be evaluated in terms of the structural characteristics of neighbouring reef systems.

Measures of Damage at the Community Level

The structural diversity of reef ecosystems is paralleled by their function a complexity. When attempting to assess damage and recovery of coral reefs in functional terms, a useful line of investigation may be the development of indices that provide a measure of total reef functions or malfunction.

One major functional characteristic of reef ecosystems is their high rate of primary production that may be 10 to 100 times greater than that of the phytoplankton in surrounding waters. Much of this reef primary production is attributable to benthic algae. A principal determinant of this high rate of productivity in nutrient poor areas is the retention and recycling of nutrients within the reef.

This feature may be a useful means of determining the status of total reef function. For example, Pilson and Betzer found that, although concentrations of phosphates in water passing over a reaf were low, there was no difference in the upstrem and downstream levels, thereby indicating that a reef recycling mechanism was present. Evidence of phosphate recycling has been found by Pomeroy *et al.*

There is also evidence of ammonium and nitrate recycling mechanisms. An increase in nutrient loss suggestive of a breakdown in nutrient recycling mechanisms may be a potential indicator of a total reef malfunction. Further evidence for the steady state nature of a reef ecosystem is provided by the ratio of gross production to community respiration (P/R) that frequently approaches unity . Again, any large be change in P/R ratio could be indicative of reef malfunction.

1. Coral Reef Recovery

Our ability to measure recovery in a coral reef system hinges on the success of a program to define or measure than is damage. Endean said that complete recovery "invoies the return of those species typically associated with corals and forming part of the coral reef community in the particular geographical region concerned and the reestablishment of the complex relationships normally existing among the species".

While this definition is theoretically correct, from a practical sense in terms of measuring recovery, it is probably too broad. Coral reefs are among the most complex marine ecosystems in terms of communities of structure and ecological function. Monitoring all aspects of reef biology in an attempt to confirm a recovery or to define a stage of recovery is not feasible.

There are, however, aspects of reef structure and function that could logically be used as indicators of a level of recovery. Specifically, a coral reef has patterns of species composition, abundance, dominance, reproduction, recruitment, growth, and mortality among the hormatypic organisms.

A gauge of recovery may be generated by comparing qualitative observations and quantitative measurements of these patterns on the recovering reef with those presumed to be characteristic of the chosen model of a fully recovered reef at the site in question. This methodology, however, requires the assumption that recovery of the coral organisms will eventually result in the recovery of the entire reef community.

Few reefs in the world have been subjectd to quantitative ecological assessments. The means that it is unlikely that any site-specific information will be available for comprative purposes. However, reef morphology and biotic community structure are typically similar over a definable range of variation within the area or region.

Therefore, there is liekly to be a basis upon which to roughly estimate the former structure and condition of nearby unimpacted reefs existing under similar conditions of natural environment, along with the remains of the impacted reef. If it is possible to infer pre-impacted structure from regional patterns, there is at least a structural and functional model for a recovered reef at the locality in question.

The measurement of recovery is complicated by the fact that recovery on coral reefs tends to be a long-term process. Endean found that the recovery period is correlated with the extent of the damages inflicted on the hard coral cover. Small scale localized destruction from natural events generally requires less than 10 years for recovery provided that major sectors of the community remain virtually intact and the area is conductive to coral growth.

Heavy destruction requires 10 to 20 years for full recovery, while especially severe impacts may require several decades for complete recovery. If a chronic source of pollution is present in the area of the damaged reef, recovery may be further prolonged or may not occur at all e.g. see Rinkevich and Loya.

Cleanup

What cleanup measures are appropriate and what are their relative effectiveness are critical questions regarding oil spills in Tundra and Taiga. Whenever possible, hydrocarbon spills should be avoided or their effects mitigated. Several measures can be employed to help accomplish this. All above ground fuel storage containers should be contained within a pit or depression lined with a hydrocarbons in previous layer. Oil pipelines should be designed to withstand weathering, corrosion, and limited displacements such as those due to earthquakes. Finally, whenever possible, oil pipelines should be routed to avoid sensitive ecological stress such as populations of endangered species.

Unfortunately, containment or avoidance is not always possible, so cleanup measures for the various soil and vegetation types must be considered. Once a spill has occurred, the choice of celanup strategies will depend upon the type of oil soiled, the nature of the spill the amount of oil, the season of the spill, and the location of the spill. Site aspect, vegetation, and soil types are critically important factors. In addition, the presence of permafrost in most areas of tundra and much of the taiga will preculude the use of many cleanup methods routinely used in warmer climates or nonpermfrost areas.

For example, in permafrost areas with a high ice content, any surface disturbance, such as removal of the oil-soaked organic soil surface horizon, tillage, or vehicle movement will induce thermal degradation of the permafrost and subsidence (thermokarast), which will increase rather than decrease total environmental damage of the site.

Under such conditions, passive (or no active) cleanup should be considered. Passive measures may also be the best means of facilitating natural recovery in cases of small spills of crude oil or contained or limited spills of industrial oils with high proportions of light fractions, much of which will be readily lost by volatilization. However, recent work by Linkins and Fetcher has shown that residual oil in the Oe-OI horizon of Tussock Tundra soils may alter reproductive and biomass allocation patterns of *Betula nana* and *Eriphorium vaginatum* such that their successful reestablishment in oil-contaminated soils is questionable.

When active cleanup measures are necessary, there is heavy reliance on site access for transporting equipment and manpower to the spill. In remvoe locations, this can involve considerable cost as well as tbe environmentally damaging unless it can be accomplished with the appropriate equipment, especially during the persnow or thawed soil times of the year or snow presses on.

Generally, the choice of which active celanup techniques are used depends upon the time of year when cleanup must occur. In winter periods when the gorund is frozen or snow covered both access and cleanup are facilitated. The dormant stage of the plants and low winter temperatures that aid in cooling and congealing crude oil so that its spread is limited both contribute to minimizing damage.

Snow also acts as an absorbent and cushion protecting surrounding vegetation from oil exposure as well as vehicle damge. Generally the oil-contaminated snow can be mechanically scraped off the site, removed, and placed in properly contained areas such as was done at the February 1978 Steele Creek spill along TAPS.

Burning the spilled hydrocarbon may also be used as a cleanup measure either before or after mechanical removal. Combustion may be difficult to initiate and generally will be incomplete. However, burning can help to remove the more toxic light fractions of the oil if done soon after the spill occurs.

Small spills may be difficult to ignite if not immediately burned, but large spills seem to burn readily even several months later. This may be due to a large proportion of the light fractions being rapped within the soil in large spills as surface weathering produces as extensive asphalitic cover. Burning is relatively inexpensive, but may leave a tarry residue on the surface that can inhibit microbial decomposition of the oil and vegetative recovery.

Finally, following any winter cleanup procedure, appropriate measures should be taken to ensure against, remobilization of the oil during snow melt. Such measures as absorbent booms, straw, or in nonpermafrost and thaw stable permafrost areas, containment ditches or dikes may be used. The hydrocarbon will then be intercepted or contained to facilitate removal.

The choice of active cleanup measures during the summer or snowfree period will depend upon the phenological stage of the plant, the soil moisture content, depth of the soil water table, and ice content of the permafrost as previously discussed.

Mechanical removal of oil contaminated soil should be considered only at times when potential damage to plants is great from prolonged extensive contact wth the hydrocarbon, the ice content of the permafrost if low, and there is relatively easy access to the spill site. Also, the volume of the spill should be such that natural recovery is unlikely and there is high potential for extensive contamination of adjacent areas through contained movement or later remobilization.

If removal of the contaminated soil is not possible, burial can be considred. However, burial should be done only in thermally stable, frozen material since subterranean remobilization can occur and cause contamination of adjacent areas.

Oil may also be removed by floating the oil and vacuuming it off the surface of the water. This was used successfully at the TAPS value 7 spill in July 1977, ready access to the site as well as availability of a large volume of water, a wet site, and feasibility of containment. However, in some cases, oil may continue to appear on the surface of the water for a year of more after the spill as it is displaced from the surface organic horizon.

Finally, manual cleanup with minimal soil disturbance should be considered in permafrost areas. Absorbents such as straw or commercial materials can remove much of the oil present on the ground surface. Although costly and very time consuming, this method will minimize any physical disturbance and decrease the chance of thermal degradation of the spill and adjacent sites.

The extensive reliance on reseeding with non-native species and long-term fertilization, whereas a relatively easy but expensive means of revegetation,does not guarantee optimal long-term revegetation or restoration. Observations of natural restoration on a 1949 disturbance and on a 1970 disturbance at Barrow, alaska suggest that it may be preferable to let some hydrocarbon disturbances recover naturally through reinvasion by native plants in lies of reseeding with exotic plants.

Recent work discussed by Shaver *et al.* that shows that the shallow organic soil horizons can serve as a major seed bank for revegetation suggests that preservation of the upper organic soil horizons may also be preferable if possible to maximize potential for native plant reestablishment. Their discussion also points out the possible deterrents of fertilization on delaying restoration of the natural vegetation.

Penetration of the hydrocrbons into the soil plant rooting zone as previously discussed will influence the choice of revegetation/restoration techniques. On wet sites where there is limited penetration of the hydrocarbon into the soil rooting zone, natural recovery by the native vegetation may be sufficient if cleanup activities adequately remove the surface hydrocarbon contamination.

On drier sites, however, where the hydrocarbon will rapidly penetrate the soil rooting zones, killing the majority of the vegetation, it may be necessary to institute extensive soil cleanup, rapid revegetation efforts. This will be especially true if site stability (othermokarst, slope erosion, etc.) is a problem. Regardless of the cleanup, revegetation/restoration activities it should be reiterated that care should be taken to preserve of site integrity. Slower spill site revegetation/restoration should always be preferable to risking both on and off site degradation.

Hydrocarbons spills in the Arctic and subarctic cause both short-and long-term effects on the soils microbial populations, vegetation, and wildlife. Although a number of descriptive studies on the effects of hydrocarbon spills have been conducted, these have generally focused on short-term effects.

Furthermore, there has been very limited development in our understanding of the important mechanisms of the direct toxic effects of oil and the effects of altered soil characteristics on organisms. When our ignorance in these areas is compounded by our limited knowledge of the in site rates and nature of hydrocarbon degradation in cold soils, it makes it very difficult to estimate effectively the duration of the influence of hydrocarbon in the soil. Likewise, it is very difficult to predict which cleanup, revegetation techniques will be most successful in alleviating hydrocarbons related deterrents to microbial population and vegetation growth.

INDEX